T0325391

Experimental Astrophysics

AAS Editor in Chief

Ethan Vishniac, Johns Hopkins University, Maryland, USA

About the program:

AAS-IOP Astronomy ebooks is the official book program of the American Astronomical Society (AAS), and aims to share in depth the most fascinating areas of astronomy, astrophysics, solar physics and planetary science. The program includes publications in the following topics:

GALAXIES AND COSMOLOGY

INTERSTELLAR MATTER AND THE LOCAL UNIVERSE

STARS AND STELLAR PHYSICS

EDUCATION, OUTREACH, AND HERITAGE

HIGH-ENERGY PHENOMENA AND FUNDAMENTAL PHYSICS

THE SUN AND THE HELIOSPHERE

THE SOLAR SYSTEM, EXOPLANETS, AND ASTROBIOLOGY

LABORATORY ASTROPHYSICS, INSTRUMENTATION, SOFTWARE, AND DATA

Books in the program range in level from short introductory texts on fast-moving areas, graduate and upper-level undergraduate textbooks, research monographs and practical handbooks.

For a complete list of published and forthcoming titles, please visit iopscience.org/books/aas.

About the American Astronomical Society

The American Astronomical Society (aas.org), established 1899, is the major organization of professional astronomers in North America. The membership (~7,000) also includes physicists, mathematicians, geologists, engineers and others whose research interests lie within the broad spectrum of subjects now comprising the contemporary astronomical sciences. The mission of the Society is to enhance and share humanity's scientific understanding of the universe.

Experimental Astrophysics

Elia Stefano Battistelli
University of Roma la Sapienza, Physics Department, Rome, Italy

IOP Publishing, Bristol, UK

ISBN   978-0-7503-3119-7 (ebook)
ISBN   978-0-7503-3117-3 (print)
ISBN   978-0-7503-3120-3 (myPrint)
ISBN   978-0-7503-3118-0 (mobi)

DOI   10.1088/2514-3433/ac0ce4

Version: 20211001

AAS–IOP Astronomy
ISSN 2514-3433 (online)
ISSN 2515-141X (print)

British Library Cataloguing-in-Publication Data: A catalogue record for this book is available from the British Library.

Published by IOP Publishing, wholly owned by The Institute of Physics, London

IOP Publishing, Temple Circus, Temple Way, Bristol, BS1 6HG, UK

US Office: IOP Publishing, Inc., 190 North Independence Mall West, Suite 601, Philadelphia, PA 19106, USA

...to Patrizio, Penelope, Miriam, and Magda

"E' così bello fissare il cielo e accorgersi di come non sia altro che un vero e proprio immenso laboratorio di fisica che si srotola sulle nostre teste."

"It is so nice to stare at the sky and realize how it is nothing more than a real immense physics laboratory that unrolls on our heads."

Margherita Hack

Contents

Preface

Astrophysics is a fascinating discipline. We are now living in the so-called multi-messenger era in which often weak and elusive astrophysical phenomena need to be studied using different and orthogonal probes and information carriers in order to be fully understood. Among the different information carriers for astrophysics that can bring new information about the cosmos, we can list: nuclear and subnuclear particles, cosmic rays, cosmic dust, meteorites, dark matter, gravitational waves, and, finally, electromagnetic waves at all different wavelengths, to which this book is mainly dedicated.

Different techniques need to be employed and developed to detect and carefully characterize electromagnetic waves arising from astrophysical objects depending mainly on their energy (frequency) and other characteristics, such as spectral features and polarization. Usually, astrophysicists need to extract a faint (astronomical) signal embedded in large interferences and noise. Thus, special techniques have to be developed and new and sensitive detectors have to be designed based on the most advanced technologies in materials science, electronics, and informatics.

This book intends to give to advanced bachelor university students a description of the most popular and advanced techniques employed in astrophysics, including the use of the most popular astronomical instrumentation. After an introduction about different "carriers" of astrophysical information, such as gravitational waves, neutrinos, and cosmic rays, this book will focus on electromagnetic radiation and its detection.

Spanning from radio astronomy to X-ray astronomy, this book will give a general description of astrophysical observables, such as flux, brightness, throughput, and magnitude. It will describe general concepts about geometrical and physical optics at the different wavelengths, in an astronomical context, including the concepts of lenses, mirrors, antennas, telescopes, the focal plane, angular resolution, the field of view, and the diffraction limit.

The theory of signals will be introduced, including the concept of transmission lines, Fourier analysis, power spectra, autocorrelation functions, the Wiener–Khinchin theorem, sampling, quantization, and the Nyquist–Shannon theorem.

The detection of an astrophysical signal is the focus of this book. Different receiving techniques will be described, such as coherent radiometers, semiconductor and superconductor thermal bolometers, charge-coupled devices (CCDs), and X-ray calorimeters.

Noise and its origin, and the signal extracted from it, will also be among the topics of this book. After an introduction to semiconductor electronics, the general concepts of (as well as the use of) analog circuits, filters, lock-in amplifiers, and operational amplifiers will be presented. After that, some general concepts about digital electronics and the use of microprocessors and field-programmable gate arrays (FPGAs) will be presented. Cryogenics and vacuum techniques will also be described in this book.

This book is intended to be a guiding text to introduce students to several astrophysical techniques, to be further detailed based on the student's interest. Also, it should give the student a foundation for carrying out laboratory activities and using the most common astronomical instrumentation and laboratory devices.

Author Biography

Elia Stefano Battistelli

 E. S. Battistelli is an *experimental astrophysicist* and *cosmologist* working on several aspects of astrophysics, including the study of the cosmic microwave background (CMB) radiation. Battistelli is currently Associate Professor in the Physics Department of the Sapienza University of Rome, where he teaches the Astrophysics Laboratory course for third-year bachelor physics students and the General Physics course for biotechnology students in the Medicine faculty. In Battistelli's research, he develops radio-frequency, microwave, and millimeter/submillimeter instrumentation for millimeter astronomy and CMB observations and data analysis. His main research interests are the study of the interaction of CMB photons with hot electron gas in clusters of galaxies, the Sunyaev–Zel'dovich (SZ) effect, the *polarization* of the CMB, including the traces that can be possibly found in it of the stochastic gravitational-wave background generated by the inflationary expansion of the universe, as well as the *spectral distortions* in the CMB frequency spectrum. Also, Battistelli is interested in the study of the *Anomalous Microwave Emission* and its origin.

Contributors

Paolo de Bernardis
Department of Physics, Sapienza University of Rome

Fabio Columbro
Department of Physics, Sapienza University of Rome

Experimental Astrophysics

Elia Stefano Battistelli

Chapter 1

Astrophysical Observables

We give here an introduction to the different kinds of astrophysical carriers, such as nuclear and subnuclear particles, cosmic rays, cosmic dust, meteorites, dark matter, and gravitational waves. We will leave electromagnetic waves for the next chapter, where a general description of astronomical optics will also be given.

1.1 Astrophysical Information Carriers

The wide majority of astrophysical measurements and observations are carried out through the detection of electromagnetic waves. This text is largely dedicated to the methods and instrumentation used for making these observations. Modern astrophysics in fact makes use of observations of celestial "bodies," the interpretation of those observations, and their comparison with physical theories that explain the observed astrophysical phenomena. However, there are several carriers of astrophysical information that give astronomers the opportunity to study astrophysical processes often orthogonally to what can be done with electromagnetic wave observations. In this sense, multimessenger astronomy[1] (when possible) allows one to combine observations made using different carriers of information and offers an enormous potential towards understanding objects that were previously indecipherable through complementary and orthogonal information. Examples are gravitational-wave observations followed (follow-up) by observations with X-ray telescopes and observations triggered by evidence of neutrino observations in search of the γ-ray electromagnetic counterpart.

Among the possible carriers of astrophysical information we can list the following:

- Cosmic rays
- Meteorites

---

[1] Multimessenger astronomy is a term that indicates coordinated observations performed with different carriers or astrophysical messengers, such as electromagnetic waves, gravitational waves, neutrinos, or cosmic rays.

doi:10.1088/2514-3433/ac0ce4ch1

- Cosmic dust
- Gravitational waves
- Neutrinos
- Dark matter
- Electromagnetic waves.

This text will introduce the reader to the different methods of making measurements on different carriers of information and what type of instrumentation is used. Particular attention will be paid to the detection techniques of electromagnetic waves and the instruments to be used. Each measuring instrument has limitations, some of a fundamental (quantum) nature, others of a practical nature. An astrophysicist (observational and/or experimental astrophysicist) takes care to select or develop instrumentation suitable for the different astrophysical observables in order to prove or refute a theory.

1.2 Cosmic Rays

Cosmic rays[2] are high-energy particles of extraterrestrial origin. The energies with which they are characterized are between[3] 10^8 eV and 10^{20} eV (~10^7 times the maximum energy achievable at the Large Hadron Collider). From a general classification, cosmic rays are formed by:

- 89% protons (hydrogen nuclei)
- 9% alpha particles (helium nuclei)
- 1% heavier nuclei
- 1% electrons, e−
- Photons, neutrinos, antimatter, etc.

In addition to the Sun, which emits low-energy cosmic rays, a significant fraction of cosmic rays comes from explosions of novae and supernovae (stellar explosions in the final phase of a star's existence), even though in part they come from active galactic nuclei, AGNs (galaxies with active nuclei that emit radiation and matter), and provide information on their physics. The average flux of cosmic rays depends on the solar wind and the terrestrial magnetic field that "filters" and conveys the cosmic rays to the polar areas.[4] Figure 1.1 shows the energy spectrum of known cosmic rays. This spectrum can be described by a power law $F \propto E^{-\alpha}$ with exponent $\alpha = 2.7$ for energies lower than 10^{15} eV, while $\alpha = 3.0$ for higher energies up to values on the order of 10^{18} eV, in which the spectrum becomes again less steep.

---

[2] The name "ray" suggests the electromagnetic nature (photons) of these particles. In reality this is only a historical confusion, in that, for the wide majority, cosmic rays are particles.
[3] 1 eV = $1.602176565 \times 10^{-19}$ J and is defined as the energy gained (or lost) by an electric charge of a single electron, which moves in a vacuum between two points between which there is a difference in electrostatic potential of 1 V.
[4] It is precisely the interaction of the solar wind particles with the ionosphere that causes the excitation and de-excitation of the atoms of the atmosphere that cause the polar auroras.

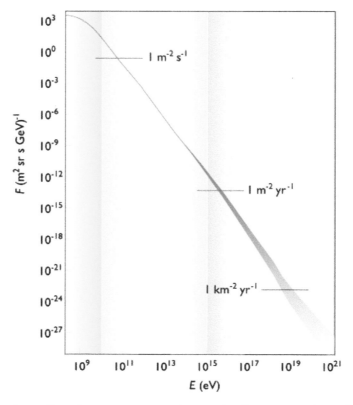

Figure 1.1. Abundance of cosmic rays per square meter, per steradian, per second, as a function of their energy. There are three regions considered to be of solar origin (yellow), of Galactic origin (blue), and of extragalactic origin (violet). Credit: Wikipedia: Sven Lafebre (CC BY-SA 3.0).

When cosmic rays (called primary cosmic rays) hit and interact with the nuclei of the molecules of the Earth's atmosphere, they trigger chain processes called showers that produce secondary particles. The secondary particles, as one gets closer to the sea level, are less energetic than the primary particles and are in turn divided into a "soft" component (30% of the total), which is composed mainly of electrons and photons and is characterized by a low penetrating power (a few centimeters) through dense media such as iron and lead, and a "hard" component (about 70%), which is mainly composed of muons that can penetrate even more than a meter. Among the most common processes is that of the decay of charged pions π into muons μ, and the production of neutrinos ν:

$$\pi^+ \rightarrow \mu^+ + \nu_\mu$$
$$\pi^- \rightarrow \mu^- + \overline{\nu}_\mu$$

(1.1)

One of the still-open questions in the study of cosmic rays is the existence of cosmic rays with energies above 10^{20} eV. Observations (that are unconfirmed) by the Japanese experiment AGASA[5] recently revealed evidence of the existence of these particles, which should have originated in nearby extragalactic sources (within 200×10^6 lt-yr), due to their interaction with the photons of the cosmic microwave background; however, the existence these sources is unknown.

The detection techniques of cosmic rays are all based on their ability to ionize materials thanks to their energy. Cherenkov telescopes exploit the Cherenkov effect in the atmosphere (air Cherenkov) that occurs when a particle travels in a medium with a speed greater than[6] c/n (c being the speed of light in a vacuum, and n the refractive index of the medium in question), ionizing the medium and producing non-symmetric dipoles. If the particle velocity is lower than c/n, each dipole relaxes elastically and there is a return to mechanical equilibrium as soon as the particle has passed. If, instead, the speed of the particle is sufficiently high compared to the response time of the medium, then the energy produced by this perturbation radiates like a coherent shock wave, forming a cone of light around the direction of propagation of the particle (Figure 1.2).

At sea level, the refractive index of air is $n = 1.00029$. For a relativistic particle with $v/c = 0.9999$, there is Cherenkov light production, and the angle of light propagation is given by:

$$\cos(\theta) = \frac{c}{n}t \times \frac{1}{\beta ct} = \frac{1}{\beta n} \rightarrow \theta \sim 1.3°. \tag{1.2}$$

The maximum development of showers (therefore, the maximum production of Cherenkov light) occurs at a height of ~10 km.

Other types of cosmic-ray detectors include scintillators, which are formed of materials that exhibit luminescence if excited with ionizing radiation and are sometimes followed by photomultipliers that amplify the signal, or bubble chambers that are composed of a liquid superheated beyond the boiling temperature, in metastable conditions, in which the particles cause the ionization of nuclei and create visible traces in the medium. The ability to use one type of particle detector or another depends on the energy of the particle one wants to reveal. The combination of different detectors that measure at different energies and also have the ability to determine the charge and speed of the particles opens up important research lines for the study of antimatter, the understanding of dark matter, and the identification of physical theories other than the standard model.

Cosmic rays have measurable effects on:
- Earth's atmosphere (e.g., they keep isotopes unstable and have weather effects)
- Humans (there are studies of their effects on pilots and astronauts)

---

[5] http://www-akeno.icrr.u-tokyo.ac.jp/AGASA/
[6] Note that the speed of light in the medium is below the speed of light in a vacuum, so in principle any particle can travel faster that this limit.

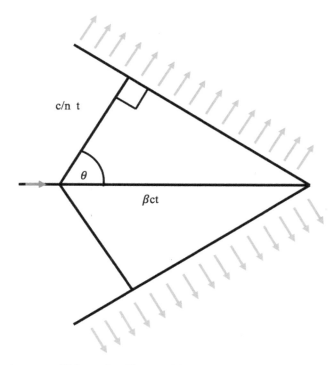

Figure 1.2. Cherenkov cone of light produced by a particle traveling at a speed higher than the speed of light in the medium.

- Detectors and digital electronics (e.g., transients in semiconductor memories and microprocessors, single event upsets, SEUs, etc.).

The last effect is of great interest for the development of astrophysical instrumentation that carries out measurements at observational sites on high mountains, on stratospheric balloons, or on satellites, along with related "radiation hardness" activities through the use of particle accelerators and the mitigation of effects.

1.3 Meteorites

A meteorite is what remains of a meteoroid after it interacts with the atmosphere (the mesosphere, at a height of 50–80 km, see Appendix A). Meteorites heat up due to the strong compression of the air in front of the meteorite and form so-called fireballs. They are extraterrestrial objects whose chemical composition can provide us with very valuable information about the cosmo. Meteorites come from our solar system and carry information about its origin. While the rocks found on Earth have undergone reprocessing and modification on Earth, meteorites can carry information about the solar system from when it originated. A small percentage of meteorites found on Earth come from the Moon or planets of the solar system, bringing with them also information about the atmospheres of the planets of the solar system (Figure 1.3).

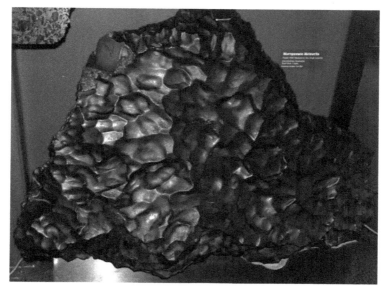

Figure 1.3. Example of an iron meteorite. Credit: Wikipedia: James St. John (CC BY 2.0).

Meteorites are historically divided according to their chemical composition:
- Rocky (silicates)
- Ferrous (iron and nickel)
- Ferro-rocky (metals + rock).

Today, they are divided into:
- Primitives: chondrites, which did not differentiate and did not undergo excessive heating, that make up 85% of the sample
- Differentiated: achondrites that originate from other celestial bodies and formed by differentiation.

The study of meteorites is linked to chemical and physical–structural analyses carried out by geologists. A fervent activity concerns the identification of real meteorites with respect to common rocks, and therefore the distinction between fallen and found meteorites, in mainly desert areas (for example, the Sahara, Oman, and Antarctica). It should not be surprising that we include Antarctica in the list of desert areas. The south pole area is in fact a desert, in the sense that the mean rainfall there is less than 200 mm of precipitable water per year, and it is all concentrated on the coast. Elements to be evaluated in the study of meteorites are their diversity with respect to the surrounding environment and their being polished on one side. The prevailing analyses are of the chemical–physical type in order to determine the characteristics and the origin of the meteorite. The dating of the meteorite is of fundamental importance: the Rubidium-87 decay in Strontium-87 is typically used, which by *beta decay*, has a half-life of 4.75×10^{10} yr.

Figure 1.4. The Horsehead Nebula: an exceptional example of optical light extinction by interstellar dust. Credit: Wikipedia: Ken Crawford (CC BY-SA 3.0).

1.4 Cosmic Dust

The cosmic dust is predominantly formed of particles ranging in size from a few molecules to a fraction of a micron. A small fraction of these particles are larger in size and are formed in stars (silicates, carbonaceous grains, graphite, and so on). Interstellar dust provides us with information on star formation. In fact, interstellar dust is a fundamental constituent of star formation and is a tracer of it. Cosmic dust absorbs stellar radiation, changes its color, and emits thermally and, possibly, as an electric dipole, due to the rapid rotation of the grains themselves. Usually, the light both absorbed (in the optical range) and emitted (in the infrared or microwave range) is observed, and due to symmetries in the arrangement of the dust grains, the study of the polarization of the observed radiation is also extremely important. Before interstellar dust was considered important in itself, the concept of extinction was known and defined (Figure 1.4). In general, blue light is absorbed more than red; therefore, extinction causes both a reduction in intensity and a change in "color" of the emission of the stars. The study of cosmic dust is therefore mainly of the classical astronomical type in the infrared and submillimeter bands. It is, however, possible to study cosmic dust also through the direct collection of dust grains.

The collection of cosmic dust is practically impossible on Earth because the stratosphere (see Appendix A) blocks the vast majority of the flux. It is therefore necessary to go into space using a satellite, or to the stratosphere using a stratospheric balloon. The collection of dust grains is mainly mechanical, which creates mechanical and physical problems depending on the type of materials used. Materials used for the entrapment of the powder include *aerogel*, a gel in which the liquid is replaced with a gas and that has, thanks to its characteristics of lightness, low density (as low as 0.01 g cm^{-3}), low thermal conductivity, and mechanical resistance, the necessary features for use in a space mission. This is, for instance, the material used in the NASA Stardust[7] satellite. Cosmic dust impacts on a satellite that collects dust while traveling at an average speed of ~10 km s^{-1}, and the mechanical characteristics of the aerogel allow for the efficient and uncontaminated collection of these particles (Figure 1.5).

1.5 Gravitational Waves

Gravitational waves are deformations of spacetime that propagate like waves: they are a prediction of general relativity and were revealed for the first time in 2015 September by the LIGO–VIRGO collaboration.[8] The production of gravitational waves is linked to an occurrence in which huge masses vary their spatial distribution (accelerating without spherical symmetry) and produce a wave front.

Gravitational wave sources are:
- Rotating neutron stars (e.g., pulsars)
- Supernova explosions
- Colliding, spiraling black holes
- Cosmic inflation.

Gravitational waves can provide complementary and orthogonal information with respect to that of electromagnetic waves. Based on this assumption, a branch of astrophysics called multimessenger astrophysics has been developed recently, in which different types of observations are used (for example, gravitational waves and

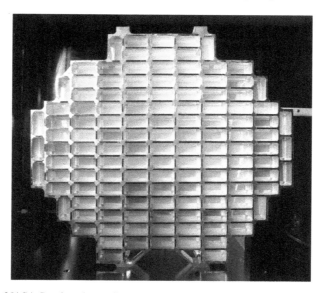

Figure 1.5. The NASA Stardust dust collector made with aerogel blocks. Credit: Courtesy of NASA.

---

[7] https://stardust.jpl.nasa.gov/home/index.html
[8] https://www.ligo.org/ and https://www.virgo-gw.eu/

electromagnetic waves) to allow for the study of astrophysical phenomena from multiple points of view, ultimately to confirm or refute theoretical predictions.

Gravitational waves can be divided into four categories:

- Continuous: these waves come from a specific direction in the celestial sphere and are caused by massive (imperfect) objects that rotate and generate gravitational waves without variation in amplitude.
- Burst: these waves originate from single, unexpected events that generate gravitational waves with strong amplitudes and that rapidly decay.
- Spiral binaries: these waves are caused by massive objects that spiral and collapse. They are also of strong amplitude and have a modulation correlated with their amplitude.
- Stochastic: these waves are generated by a multitude of objects that add up their contributions in a statistically inconsistent way. Another mechanism of generation of stochastic gravitational waves is that of cosmic inflation, a prediction of the standard cosmological model, which proposes an exponential expansion of the universe when it was only $\sim 10^{-35}$ s old and which would be at the origin of a stochastic background of gravitational waves that would be very difficult to reveal directly.

On 2016 February 11, LIGO announced the first detection of an event like the collision (spiraling) of two black holes with ~30 times the mass of the Sun. This earned Rainer Weiss, Barry C. Barish, and Kip S. Thorne the Nobel Prize in Physics in 2017 for their contribution to the realization of the US observatory LIGO, which made possible the first direct detection of gravitational waves. On 2017 August 17, the LIGO and VIRGO experiments observed another event that was in some ways even more important: thanks to triangulation, made possible precisely by the presence of multiple detectors in different positions on Earth, they observed an event and were able to determine the direction from which this signal came. The gravitational-wave signal lasted for ~100 s. This allowed observers from all over the world (and with satellites) to observe a short γ-ray burst using the Fermi[9] and INTEGRAL[10] space telescopes. The optical counterpart was observed in the galaxy NGC 4993 about 10 hr later using the Chilean SWOPE telescope[11] in the near-infrared. The gravitational wave and the electromagnetic-wave signal (in total, 70 observers across the world observed the event) allowed researchers to determine that at the origin of this event there was a collision of two neutron stars with a total mass of several solar masses. The collision would have been followed by the explosion of a *kilonova*, an event that is of fundamental importance to the formation of chemical elements heavier than iron. Since these first events, several others have occurred, allowing us to enter the multimessenger era.

The detection of gravitational waves has been a fervent activity since the 1960s. The first detectors made use of large metal bars in which a gravitational wave should

---

[9] https://fermi.gsfc.nasa.gov/
[10] https://sci.esa.int/web/integral
[11] https://obs.carnegiescience.edu/swope

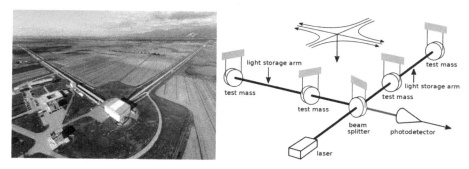

Figure 1.6. Left: a photo of the VIRGO interferometer in Cascina (Pisa, Italy). VIRGO is an international collaboration funded by the French CNRS (http://www.cnrs.fr/) and the Italian INFN (http://home.infn.it/en/). Credit: The Virgo collaboration (CC0). Right: a conceptual scheme of the LIGO interferometer. LIGO is an American experiment funded by the NSF (https://www.nsf.gov/). Credit: Ligo.gif (Public Domain).

vary the resonance frequency. Modern detectors like LIGO and VIRGO are laser interferometers (Figure 1.6). They allow for the measurement of the differential variation, caused by the passage of a gravitational wave, in the lengths of the two arms of the interferometers, and they monitor the effect through the constructive or destructive interference of the laser inside them. Although the measurement is conceptually simple, the main difficulties are technical and related to sensitivity. The effect of a gravitational wave is in fact very weak (the strongest waves are estimated to produce a differential deformation of 10^{-18} m), and it is therefore necessary to produce cryogenic vacuum equipment, isolated from any external disturbances. LIGO and VIRGO are interferometers with arms of several kilometers under vacuum, seismically superinsulated with active and passive damping. Passive insulation is obtained, for example, with suspended optics (e.g., glass cables). Active insulations, however, are developed by placing all of the experimental apparatus on active platforms that counteract any earth movements.

1.6 Neutrinos

Neutrinos (ν) are subnuclear particles of very small mass and neutral charge. They are generated by decays (e.g., β) or nuclear reactions. Following the standard model of elementary particles, each lepton (electron e, muon μ, and tauon τ) has a neutrino associated with it (ν_e, ν_μ, ν_τ). Examples sources of neutrinos are:

- The fusion of hydrogen, which produces ν_e
- The Sun, which produces solar neutrinos
- Supernova explosions (and the production of neutron stars), which are associated with neutrino bursts
- A cosmic background that is foreseen by the standard cosmological model. Following the expansion of the universe, the neutrino background would have decoupled from the rest of the components in the universe, producing today a 1.9 K cosmic neutrino background.

Their low interacting nature poses obvious problems for their detection. That said, from an astronomical point of view, they can help provide information about celestial bodies that are too distant to detect using photons. Furthermore, neutrino properties allow scientists to discover new phenomena. Important examples are the fact that many dark-matter candidates emit neutrinos, and the fact that cosmic rays of very high energy are associated with the production of neutrinos.

Until 2017, the extraterrestrial evidence of neutrinos was limited to solar neutrinos, the supernova SN1987A, and three very energetic nonlocalized events revealed by the IceCube experiment[12], which were not confirmed by follow-up observations. The emission of neutrinos is associated with super-energetic events that produce cosmic rays with energies of, for example, 10^{20} eV. That said, charged particles interact a lot with matter and magnetic fields and are therefore easily absorbed. Furthermore, energetic particles with energies above 10^{20} eV are still absorbed by the photons of the cosmic microwave background.

From an observational point of view, the search for extragalactic neutrinos is extremely difficult due both to the weakness of the signal and to contaminants. One should just think that the enormous flow from the Sun (which is enormous compared to that expected from outside the solar system) produces only one interaction for every 10^{36} target atoms. In order to detect extra solar system neutrinos, it is necessary to employ powerful shielding systems, such as an entire mountain or even the Earth itself. In addition, huge masses are needed to produce signals. In this case, huge masses of water (at sea) or ice from the Antarctic pack are used. It is necessary to be able to determine the direction of the possible neutrino to identify the events that come, for example, from the other side of the Earth. One of the typical processes is due to the interaction of the neutrino with matter (protons, p), creating muons (and neutrons, n):

$$\overline{\nu}_\mu + p \rightarrow \mu^+ + n. \tag{1.3}$$

One of the most successful neutrino detectors is the IceCube experiment (Figure 1.7). IceCube is a neutrino "telescope" optimized to detect neutrinos with energies on the order of a TeV. It reveals high-energy astrophysical neutrinos that move upwards, after crossing the Earth, and that produce muons μ (and very rarely tauons τ) that interact with the ice. The muons emit Cherenkov light, which is revealed by 5160 detectors located within a volume of one cubic kilometer, between 1500 m and 2500 m below the ice surface in 86 strings dropped into holes in the ice.

In 2017 September, the IceCube experiment disclosed an alert to the global astronomical community regarding an event that was possibly consistent with the passage of a highly energetic extragalactic neutrino. Roughly 20 telescopes from around the world turned their attention to the constellation Orion. The Fermi γ-ray satellite was the first to observe a γ-ray flare consistent with the presence of a blazar (originating from an elliptical galaxy with a supermassive black hole inside that

---

[12] http://icecube.wisc.edu/

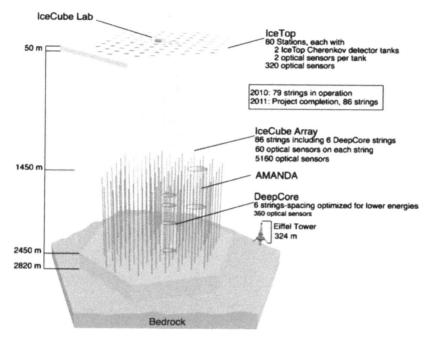

Figure 1.7. IceCube is a neutrino observatory installed at the Amunsen Scott base at the south pole in Antarctica. IceCube is a cubic kilometer detector inside the ice between 1500 m and 2500 m below the surface, with 5160 detectors distributed over 86 strings. Credit: Wikipedia: Nasa-verve (CC BY 3.0) IceCube Science Team—Francis Halzen, Department of Physics, University of Wisconsin.

occasionally emits its jets towards the Earth). Subsequently, the MAGIC experiment[13] observed the same impulse. The blazar **TXS 0506+056** offered the first unequivocal detection of extragalactic neutrinos and opened important frontiers of the multimessenger study of ultra-energy sources in our universe.

1.7 Dark Matter

The mass-energy of our universe is made up of about 5% "ordinary" matter (astrophysicists call it baryonic, although it is formed also by leptons), ~27% dark matter, and ~68% dark energy. Dark matter does not emit electromagnetic radiation and manifests itself only through gravitational interactions. Its detection is clearly an indirect detection (Figure 1.8).

The evidence (which is mainly astrophysical) for dark matter is on different scales:

- When measuring the rotation curves of different spiral galaxies, we detect a speed as a function of distance from the nucleus that is greater than we would expect if the galaxy was formed of the only matter that we can see electromagnetically.

---

[13] http://www.magic.iac.es/

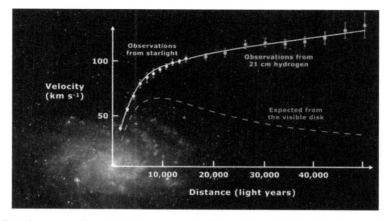

Figure 1.8. Rotation curve of a spiral galaxy measured with starlight and with the hydrogen 21 cm line. The measured speeds, as a function of the radius, can only be explained thanks to the presence of a halo of dark matter that pervades the galaxy. Credit: Wikipedia: Mario De Leo (CC BY-SA 4.0).

- When observing the motion of galaxies within clusters of galaxies, we see a higher speed than would be expected if the mass was solely due to visible matter.
- The mass budget that comes from analyses that take into account the propagation of cosmic microwave background photons on the scale of the whole universe foresees the existence of nonvisible matter that interacts gravitationally but not electromagnetically.
- Gravitational lensing implies the presence of dark matter. General relativity requires that spacetime, and therefore also the propagation of an electro-magnetic wave, be deflected due to the presence of a sufficiently large mass. Observations of electromagnetic sources, in which a massive object is placed between the source and the observer, allow for the determination of the mass of the object acting as a lens. The mass estimates obtained through gravita-tional lenses indicate the presence of a mass that has gravitational but not electromagnetic effects: dark matter.

Dark matter candidates are of different types. Examples include:
- Massive neutrinos (possibly sterile; that is, they do not interact except with gravity)
- Axions (particles proposed to explain the nonviolation of CP)
- WIMPs (weakly interactive massive particles, which do not interact via the electromagnetic force but only via the weak force)
- MACHOs (massive astrophysical compact halo objects, which are massive astrophysical objects such as neutron stars, black holes, and brown dwarfs that could explain, at least in part, the evidence of dark matter)
- The Kaluza–Klein theory (a precursor of string theory), which predicts a fifth dimension where other particles would be visible, partners of the particles of the standard model.

As already mentioned, a method for studying dark matter is through indirect studies of the Cosmic Background Radiation or through the study of density fluctuations of ordinary matter on a large scale (baryonic acoustic oscillations, BAO). An alternative method would be to produce dark matter in particle accelerators. However, there are also direct dark-matter detection methods whose effectiveness depends on the type of particle hypothesized and the energies involved. For example, WIMPs can mechanically interact with nuclei by releasing energy. To detect their presence, detectors with large volumes, large screens (as big as a mountain), and cryogenic sensors are needed. For such applications, there is a great use of pure radio lead screens. Assuming a uniform distribution of dark-matter particles, an annual modulation is expected due to the motion of the Earth around the Sun, which is added to that of the solar system around our Galaxy. For a uniform distribution, in fact, the Earth is expected to be crossed by a greater flow of dark-matter particles around 2 June (when its speed is added to that of the solar system in the reference frame of the Galaxy) and by a smaller flow around 2 December (when the two speeds cancel each other). An important result was obtained from the DAMA experiment[14], which actually observed this modulation, but was not confirmed by other experiments.

---

[14] http://people.roma2.infn.it/~dama/web/home.html

Chapter 2

Astronomical Optics

In this chapter the general concepts of optics, focused on their astronomical use, are given. Also described are the concepts of luminosity, flux, brightness, magnitude, throughput, and spectral bands. Geometrical optics, waves in a vacuum, waves in a medium, Snell's law, and the refractive index as a function of wavelength will also be presented. A worked example for the calculation of limiting magnitude will be given. Also, the basic principles of physical optics and how they apply to astronomy will be presented; particular attention will be given to the problems of diffraction and interference and the problem of the diffraction limit in astronomical observations.

2.1 Electromagnetic Waves

The origin and propagation of an electromagnetic wave can be described with Maxwell's equations. In an isotropic and unlimited dielectric of dielectric constant ε and magnetic permeability μ, in which the localized charges and the currents are zero, Maxwell's equations describe the behavior of the electric field E and magnetic field B:

$$
\begin{aligned}
\nabla \cdot \boldsymbol{E} &= 0 \\
\nabla \cdot \boldsymbol{B} &= 0 \\
\nabla \times \boldsymbol{E} &= -\frac{\partial \boldsymbol{B}}{\partial t} \\
\nabla \times \boldsymbol{B} &= \varepsilon\mu\frac{\partial \boldsymbol{E}}{\partial t}.
\end{aligned}
\tag{2.1}
$$

By calculating the rotor of the third equation above, taking into account the first equation and remembering that

$$
\nabla \times (\nabla \times \boldsymbol{E}) = -\nabla^2 \boldsymbol{E} + \nabla(\nabla \cdot \mathbf{E})
\tag{2.2}
$$

we get:

$$\nabla^2 \boldsymbol{E} = -\nabla \times \frac{\partial \boldsymbol{B}}{\partial t} = \frac{\partial (\nabla \times \boldsymbol{B})}{\partial t} = \varepsilon\mu \frac{\partial^2 \boldsymbol{E}}{\partial t^2} \qquad (2.3)$$

and so

$$\nabla^2 \boldsymbol{E} - \varepsilon\mu \frac{\partial^2 \boldsymbol{E}}{\partial t^2} = 0$$
$$\nabla^2 \boldsymbol{B} - \varepsilon\mu \frac{\partial^2 \boldsymbol{B}}{\partial t^2} = 0. \qquad (2.4)$$

From these equations, considering the dependence of the field from the temporal and spatial coordinates, the propagation equation of an electromagnetic wave with speed $\varepsilon\mu = 1/c^2$ is obtained, where c is the speed of the wave in the medium. Considering, for example, the electric field E, we have the solution:

$$E = E_0 e^{[i(k \cdot r - \omega t + \varphi)]} \qquad (2.5)$$

where φ is the initial phase, k is the wavevector, and ω is the pulsation. The wavevector k is oriented in the direction of propagation of the wave and is a function of the wavelength λ. It is related to the pulsation ω, the period T, and the frequency ν. We have the following relationships:

$$k = \frac{2\pi}{\lambda} \leftrightarrow \lambda = \frac{c}{\nu} \leftrightarrow \omega = 2\pi\nu = \frac{2\pi}{T}. \qquad (2.6)$$

In a propagating electromagnetic wave, the electric field E and magnetic field B are always perpendicular to each other and perpendicular to the direction of propagation. Assuming a source is at infinity, the wave front (i.e., the locus of points where the argument of the function is constant) is flat; that is, for a wave that propagates along the x-axis, the partial derivatives along y and z are null. In this case we speak of a plane wave, the solution of the following equation:

$$\frac{\partial^2 E}{\partial x^2} = \varepsilon\mu \frac{\partial^2 E}{\partial t^2}. \qquad (2.7)$$

If the source is localized, the wave front is spherical. In this case we speak of a spherical wave. Passing into spherical coordinates, we have:

$$\frac{1}{r^2} \frac{\partial}{\partial r} \left(r^2 \frac{\partial E}{\partial r} \right) = \varepsilon\mu \frac{\partial^2 E}{\partial t^2} \qquad (2.8)$$

from which:

$$\frac{\partial^2 (rE)}{\partial r^2} = \varepsilon\mu \frac{\partial^2 E}{\partial t^2} \qquad (2.9)$$

where the argument in the derivatives is rE. From this equation, it is evident that the electric field E propagates as $1/r$ while the power ($\propto E^2$) propagates as $1/r^2$. This is an evident consequence of energy conservation.

Electromagnetic waves have different effects and uses according to their energy, wavelength λ, or frequency:

- Radio waves (3×10^6 m $> \lambda > 0.3$ m): waves of these wavelengths are used for radio signals, TV, and mobile telephony. From an astrophysical point of view, radio astronomy deals with their detection. The branch of radio interferometry is highly developed.
- Microwaves (0.3 m $> \lambda > 3 \times 10^{-3}$ m): at these wavelengths there are transmissions for telecommunications, geopositioning systems (GPS), and radar. Of astrophysical interest, in microwaves there are important rotational line emissions of the water molecule, interstellar dust emission, and the emission of the cosmic microwave background (CMB).
- Millimeter (mm) and submillimeter wavelengths (3×10^{-3} m $> \lambda > 5 \times 10^{-4}$ m): This wavelength range is used to study star-forming regions, the CMB, and the interaction of the CMB with the galaxy clusters.
- Infrared (IR, 5×10^{-4} m $> \lambda > 0.79 \times 10^{-6}$ m): in the infrared is the emission of celestial bodies at temperatures from tens to hundreds of Kelvins. The emission of interstellar dust and the study of molecular vibrational excitations in gases are studied in IR astronomy.
- Visible (0.79×10^{-6} m $> \lambda > 0.38 \times 10^{-6}$ m): in the visible there is emission due to the quantum mechanical excitation of atomic transitions, in addition to emissions from stars.
- Ultraviolet (UV, 0.38×10^{-6} m $> \lambda > 6 \times 10^{-10}$ m): at UV wavelengths, the innermost electrons in atoms are excited and ionization begins. UV astronomy is complementary to visible astronomy.
- X-rays (6×10^{-10} m $> \lambda > 6 \times 10^{-12}$ m): X-rays are ionizing and penetrating. Of greatest interest is the Bremsstrahlung emission caused, for example, by relativistic electrons in the intergalactic medium of galaxy clusters.
- γ-rays (6×10^{-12} m $> \lambda$): γ-rays are produced in nuclear reactions. In astrophysics, γ-ray bursts are extremely interesting; they are γ-ray bursts of very high energy, probably linked to the growth of matter on a black hole.

In astrophysics, the use of electromagnetic-wave detections to observe the celestial sphere is exploited in different ways and with different "products" and methodologies:

- Imaging: the making of maps by measuring the brightness within an area of the sky.
- Photometry: the detailed measurement of the flux emitted by astrophysical sources.
- Astrometry: the measurement of the position of and distance to celestial objects.
- Spectroscopy (line and continuum): this is the measurement of the emission of astrophysical objects as a function of wavelength (a spectrum). In this case, we distinguish the study of the emission or absorption of spectral lines that are associated with molecular or atomic transitions (for which high spectral resolution is required) from spectroscopy aimed at determining the emission spectra, and the

mechanism of emission of astrophysical sources as a function of wavelength (a spectral energy distribution, SED, for which low angular resolution is required).

- Polarimetry: this is the measure of the degree and direction of polarization of an electromagnetic wave emitted by an astrophysical source. Polarization is associated with the plane of oscillation of the electric field and magnetic field associated with an electromagnetic wave and carries within it information about how electromagnetic waves are emitted.
- The study of temporal variations: some astrophysical phenomena are characterized by strong temporal variations whose monitoring allows for the extraction of information about energy emission mechanisms and the astrophysics that governs these emissions.

Above, we listed some examples of different observational characteristics and techniques, not necessarily separate from each other, that require different observational devices and techniques, and different astronomical instruments.

2.2 Observables

2.2.1 Flux, Brightness, and Magnitude

The luminosity of an object is the total energy emitted by the object itself per unit of time. The power emitted, in SI units, is measured in $W = J\ s^{-1}$. It is an intrinsic quantity of an astronomical object independent of its distance from the observer and from the experimental apparatus used to observe it. Luminosity L can be different frequency by frequency (we will see after the spectral densities). The brightness of a star clearly depends on its size and temperature (and, as you know, from its evolutionary state, spectral class, type, etc.).

Flux is the amount of energy emitted by an object per unit of time, per unit of surface, and in SI units is measured in $J\ s^{-1}\ m^{-2}$:

$$F = \frac{dW}{dA} = h\nu\frac{dn}{dt\,dA} \rightarrow L = \int F\,dA = \int dW = W_{\text{tot}} \tag{2.10}$$

where dn/dt is the photon (numeric) flux, useful only if you know the wavelength of the photons in question. In the case of an isotropic source (which emits energy uniformly in all directions), the flux is inversely proportional to the square of the distance. In fact, the same energy passes through an area at a distance r, $2r$, $3r$...

$$F(r) \cdot 4\pi r^2 = L \rightarrow F(r) \propto \frac{1}{r^2}. \tag{2.11}$$

Remember that 4π is the solid angle (in steradians) of a sphere.

Brightness is the amount of energy emitted by an object per unit of time per unit of surface, and per unit of solid angle (around a predetermined direction θ). In SI units, it is measured in $J\ s^{-1}\ m^{-2}\ sr^{-1}$:

$$B(\vec{\theta}) = \frac{dW}{dA\,d\Omega} = h\nu\frac{dn}{dt\,dA\,d\Omega} \tag{2.12}$$

which retains information of the direction of origin. In the case of a point source at a great distance, this information is superfluous, but for extended sources it is instead fundamental. Measurement units, in addition to SI units, are degrees Kelvin and the Jansky (Jy, see Section 2.2.2), for example, antenna temperature (K), MJy sr^{-1}, and counts, and their use depends on the bands and energies involved.

Brightness can be expressed in linear or logarithmic (magnitude) units:

$$\mu(\vec{\theta}) = -2.5 \log [B(\vec{\theta})] + C \qquad (2.13)$$

where C is a constant.

The apparent magnitude m derives directly from the flux (therefore, it is linked to the radiation that is measured, not that is emitted) in the following way:

$$m = -2.5 \log F + C. \qquad (2.14)$$

The minus sign, the factor of 2.5, and the logarithmic scale are present for historical/anthropic reasons. This scale is in fact linked to the classification given by Hipparchus (190–120 BC) who classified the stars visible to the naked eye in six classes from the brightest ($m = 0$, Vega) to the least bright ($m = 6$). In modern times, the staircase has clearly become richer.

Magnitudes in the solar system, for instance, are: $m_{Sun} = -26.74$, $m_{full\ moon} = -12.6$, $m_{Venus} \sim -4$, $m_{Mars} \sim -2$, and $m_{Saturn} \sim 0.6$. The γ-ray burst GRB080319B, observed on 19 March 2008 from a distance of 7.5 billion light years, had $m = 5.6$.

The difference between two magnitudes is:

$$m_1 - m_2 = -2.5 \log F_1 + C + 2.5 \log F_2 - C = -2.5 \log \frac{F_1}{F_2}. \qquad (2.15)$$

Absolute magnitude, however, is an intrinsic property of an object and derives from its luminosity. It is defined as the apparent magnitude that an object would have if it was at a distance of 10 pc (32.6 lt-yr):

$$M = -2.5 \log F_{10pc} + C = -2.5 \log \left(\frac{L}{4\pi(10pc)^2} \right) + C. \qquad (2.16)$$

The following relations will therefore apply:

- Difference between two absolute magnitudes,

$$M_1 - M_2 = -2.5 \log \left(\frac{L_1}{L_2} \right) \qquad (2.17)$$

- Modulus distance,

$$m - M = -2.5 \log \left(\frac{L}{4\pi d^2} \frac{4\pi(10pc)^2}{L} \right) = 5 \log \left(\frac{d}{10} \right) = 5 \log(d) - 5. \qquad (2.18)$$

From an astrophysical point of view, we are clearly willing to determine the intrinsic characteristics of objects, such as the brightness or absolute magnitude, to study the "astrophysics" of objects. From an observational or experimental point of view, however, we always refer to apparent magnitudes and flux because these are the quantities that we can measure. An astrophysicist observes the flux and not the luminosity.

Working example: limiting magnitude

The faintest magnitude perceptible to the naked eye is $m_0 = 6$, which corresponds to a given power limit P_{lim}. The naked eye (adapted to the dark) has a pupil with radius $r \sim 3.5$ mm ($d = 7$ mm). The use of larger objectives and telescopes of radius R increases the faintest observable magnitude to the limit m_t.

If the naked eye limiting magnitude is $m_0 = 6$, then we want to determine the limiting magnitude m_t that we can observe with a telescope of radius R and diameter $D = 2R$.

So we have:

$$m_0 = -2.5 \log\left(\frac{P_{lim}}{\pi r^2}\right) + C$$

$$m_t = -2.5 \log\left(\frac{P_{lim}}{\pi R^2}\right) + C.$$

Their difference is thus:

$$m_t - m_0 = -2.5 \log\left(\frac{P_{lim}\, \pi r^2}{P_{lim}\, \pi R^2}\right)$$

so

$$m_t - m_0 = -5 \log\left(\frac{D}{d}\right)$$

and thus

$$m_t = 6 + 5 \log\left(\frac{1}{d}\right) + 5 \log(D) = 16.8 + 5 \log(D)$$

where D is the diameter of the telescope. This is only a rough calculation that does not consider the seeing, the meteorological conditions, light pollution, etc. Also, this limit is only for observations undertaken with a telescope and with the eye placed at the ocular. It does not apply when using a CCD.

2.2.2 Spectral Densities

If the radiation detected by an experimental apparatus is not monochromatic, the flux and brightness per unit frequency (spectral densities) can be defined. We have:

$$F_\nu = \frac{dW}{dA\, d\nu} [\text{Wm}^{-2}\,\text{Hz}^{-1}] \tag{2.19}$$

$$B_\nu = \frac{dW}{dA \, d\Omega d\nu} [\text{Wm}^{-2} \, \text{sr}^{-1} \, \text{Hz}^{-1}]. \tag{2.20}$$

For wavelength units ($\lambda = c/\nu$):

$$F_\lambda = \frac{dW}{dA \, d\lambda} [\text{Wm}^{-2} \, \text{m}^{-1}] \tag{2.21}$$

$$B_\lambda = \frac{dW}{dA \, d\Omega d\lambda} [\text{Wm}^{-2} \, \text{sr}^{-1} \, \text{m}^{-1}]. \tag{2.22}$$

And per unit wave number ($\sigma = 1/\lambda$):

$$F_\sigma = \frac{dW}{dA \, d\sigma} [\text{Wm}^{-2} \, \text{m}^{1}] \tag{2.23}$$

$$B_\sigma = \frac{dW}{dA \, d\Omega \, d\sigma} [\text{Wm}^{-2} \, \text{sr}^{-1} \, \text{m}^{1}]. \tag{2.24}$$

The integrated quantities are:

$$F = \int_0^\infty \frac{dW}{dA \, d\lambda} d\lambda = \int_0^\infty F_\lambda d\lambda = \int_0^\infty F_\nu d\nu = \int_0^\infty F_\sigma d\sigma. \tag{2.25}$$

Given that

$$\lambda = \frac{c}{\nu} \rightarrow |\, d\lambda \,| = \left| \frac{c}{\nu^2} d\nu \right| \tag{2.26}$$

we have:

$$\lambda F_\lambda = \lambda \frac{dW}{dA \, d\lambda} = \lambda \frac{\nu^2}{c} \frac{dW}{dA \, d\nu} = \nu F_\nu = \sigma F_\sigma. \tag{2.27}$$

Another unit used to express the flux density is the Jansky (Jy), which is widely used in radio astronomy (Figure 2.1). A Jansky is 10^{-26} Watts per square meter per hertz:

$$1\text{Jy} \equiv 10^{-26} \, \text{Wm}^{-2} \, \text{Hz}^{-1}. \tag{2.28}$$

2.2.3 Photometric Bands

A photometer is an instrument that measures electromagnetic radiation from an astrophysical object. It measures astrophysical radiation within a spectral band by integrating over it. We can give an indication of the efficiency of a photometer by providing the maximum frequency and the FWHM of its response, but a quantitative interpretation of the data can only be obtained by knowing the efficiency $e(\nu)$; (at all frequencies, possibly by characterizing it with a spectrometer). It is possible to define the effective frequency:

Figure 2.1. Karl Jansky was an American physicist. He discovered that the Milky Way emits radio waves. The flux unit Jansky, Jy, named after him, is widely used in radio astronomy. Credit: Wikipedia: Uploaded by Saber1983 (CC BY-SA 3.0).

$$\nu_{eff} = \frac{\int_0^\infty \nu F(\nu)e(\nu)d\nu}{\int_0^\infty F(\nu)e(\nu)d\nu} \tag{2.29}$$

which represents an average of the frequencies within the filter weighted by the efficiency and the (specific) flux of the source. The weight of the source will cause the effective frequency to approach the value at which the source is most intense (e.g., where there is an intense line). The effective frequency depends on the shape but not on the amplitude of the source spectrum.

Radio Bands
In the radio band (3 Hz to 1 GHz) there is a lot of telecommunications activity, and radio astronomy is a minority. In the following is shown the International Telecommunication Union (ITU) division of the radio-frequency electromagnetic spectrum into bands, and their nomenclature (Table 2.1).

Microwave Bands
The microwave range of frequencies is also divided into bands by the American Institute of Electrical and Electronics Engineers (IEEE) and the Radio Society of Great Britain (RSGB). The classification is shown in Table 2.2.

Table 2.1. Radio Bands

Band Name	Abbreviation	Frequency	Example Uses
Extremely Low Frequency	ELF	3–30 Hz	Submarine communication
Super Low Frequency	SLF	30–300 Hz	Submarine communication
Ultra Low Frequency	ULF	300–3000 Hz	Submarine communications, communication within mines
Very Low Frequency	VLF	3–30 kHz	Navigation, time signals, submarine communication, wireless heart rate monitors, geophysics
Low Frequency	LF	30–300 kHz	Navigation, time signals, AM long-wave broadcasting (Europe and parts of Asia), amateur radio
Medium Frequency	MF	300–3000 kHz	AM(medium-wave) broadcasts, amateur radio
High Frequency	HF	3–30 MHz	Short-wave broadcasts, amateur radio, aviation communications, radar, radio communications, marine and mobile radio telephony
Very High Frequency	VHF	30–300 MHz	FM broadcasts, television, line-of-sight, ground-to-aircraft, aircraft-to-aircraft, land mobile and maritime mobile communications, amateur radio, weather radio
Ultra High Frequency	UHF	300–3000 MHz	Television broadcasts, microwave oven, microwave devices/communications, radio astronomy, mobile phones, wireless LAN, Bluetooth, GPS, amateur radio, satellite radio, remote control systems
Super High Frequency	SHF	3–30 GHz	Radio astronomy, microwave devices/ communications, wireless LAN, DSRC, most modern radar systems, communications satellites, cable and satellite television broadcasting, amateur radio, satellite radio
Extremely High Frequency	EHT	30–300 GHz	Radio astronomy, high-frequency microwave radio relay, microwave remote sensing, amateur radio, millimeter-wave scanners, wireless LAN
Tremendously High Frequency	THF	300–3000 GHz	Experimental medical imaging to replace X-rays, ultrafast molecular dynamics, condensed-matter physics, terahertz time-domain spectroscopy, terahertz computing/ communications, remote sensing

Table 2.2. Microwave Bands

Band	Frequency	Example Uses
L band	1–2 GHz	Military telemetry, GPS, mobile phones (GSM), amateur radio
S band	2–4 GHz	Weather radar, surface ship radar, some communications satellites, microwave ovens, microwave devices/communications, radio astronomy, mobile phones, wireless LAN, Bluetooth, GPS, amateur radio
C band	4–8 GHz	Long-distance radio telecommunications
X band	8–12 GHz	Satellite communications, radar, terrestrial broadband, space communications, amateur radio, molecular rotational spectroscopy
K$_u$ band	12–18 GHz	Satellite communications, molecular rotational spectroscopy
K band	18–26.5 GHz	Radar, satellite communications, astronomical observations, automotive radar, molecular rotational spectroscopy
K$_a$ band	26.5–40 GHz	Satellite communications, molecular rotational spectroscopy
Q band	33–50 GHz	Satellite communications, terrestrial microwave communications, radio astronomy, automotive radar, molecular rotational spectroscopy
U band	40–60 GHz	
V band	50–75 GHz	Millimeter-wave radar research, molecular rotational spectroscopy, and other kinds of scientific research
W band	75–110 GHz	Satellite communications, millimeter-wave radar research, military radar targeting and tracking applications, and some non-military applications, automotive radar
F band	90–140 GHz	SHF transmissions: radio astronomy, microwave devices/communications, wireless LAN, most modern radar systems, communications satellites, satellite television broadcasting, amateur radio
D band	110–170 GHz	EHF transmissions: radio astronomy, high-frequency microwave radio relay, microwave remote sensing, amateur radio, directed-energy weapons, millimeter-wave scanners

IR, Optical, and UV Bands

In optical and near-infrared astronomy, filter standards have been defined (photometric systems). This was initially defined as the U–B–V (Ultraviolet–Blue–Visual) system, also known as the Johnson or Johnson–Morgan system (Figure 2.2). Later it was extended to the U–B–V–R–I (Ultraviolet–Blue–Visual–Red–Infrared) system.

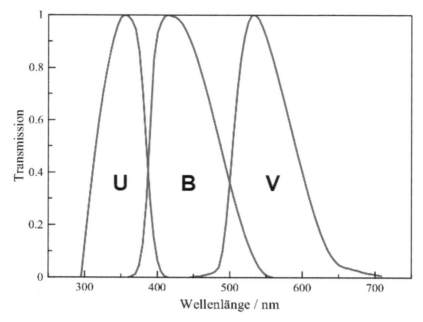

Figure 2.2. UBV photometric system. Credit: Wikipedia: By Michael Oestreicher (CC BY-SA 4.0).

Today there exist around 200 standard filter systems. In Table 2.3 a complete set of filters at infrared, visible, and ultraviolet wavelengths is given.

These filters attempt to trace the responses of the human eye (V) or a photographic plate (B), or to match the atmospheric transmission windows (see Appendix A). The filters used in practice are not ideal; therefore, they must be characterized. Also, in the definition of magnitude, we have to consider the spectral transmission of the filters. We therefore have that if:

$$m = -2.5 \log F + C \qquad (2.30)$$

then in a photometric band x:

$$m_x = -2.5 \log \int F(\lambda)e_x(\lambda)d\lambda + C_x \qquad (2.31)$$

with $m_U = U$, $m_B = B$, and $m_V = V$. The calibration standard originally defined the magnitude system in such a way that the star Vega has magnitude = 0 in all spectral bands: $U_{\text{Vega}} = B_{\text{Vega}} = V_{\text{Vega}} = \ldots = 0$. Now it includes the atmospheric extinction as well.

The difference in apparent magnitudes in different bands depends on the spectral efficiency of the bands:

$$m_x - m_y = -2.5 \frac{\log \int F(\lambda)e_x(\lambda)d\lambda}{\log \int F(\lambda)e_y(\lambda)d\lambda}. \qquad (2.32)$$

Table 2.3. UV, Optical, and IR Bands Used in Astronomy

Filter Band	Effective Wavelength (nm)	FWHM (nm)
Ultraviolet		
U	365	66
Visible		
B	445	94
V	551	88
G		
R	658	138
Near-infrared		
I	806	149
Z		
Y	1020	120
J	1220	213
H	1630	307
K	2190	390
L	3450	472
Mid-infrared		
M	4750	460
N		
Q		

The difference in magnitude in different spectral bands is defined as a color index, which is of obvious interest for the determination of stellar spectral class and temperature. We have:

$$m_B - m_V = -2.5 \frac{\log \int F(\lambda) e_B(\lambda) d\lambda}{\log \int F(\lambda) e_V(\lambda) d\lambda}. \tag{2.33}$$

A definition that should be noted down is the bolometric magnitude: the magnitude associated with the flux emitted at all wavelengths m_{bol}. It is a definition that makes sense for stellar radiation, but it should be corrected for the instrument pass band, atmospheric absorption, and interstellar extinction. It is therefore very difficult to determine and should be based upon modeling. Also, many sources have emissions related to different mechanisms in different wavelengths, which should be accounted for.

2.2.4 Throughput, Etendue, and $A\Omega$

A photometer is an instrument capable of measuring the radiative power coming from an astrophysical object within a frequency band. It includes:
- An optical system that collects radiation
- A detector that transforms radiative power usually into a measurable and recordable electrical signal

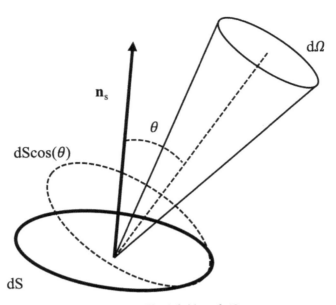

Figure 2.3. The definition of $A\Omega$.

Given a surface dS with normal vector $\mathbf{n}_s$, we can define a propagation direction forming an angle θ with respect to dS and subtending a solid angle $d\Omega$. We can define the $A\Omega$, or throughput, or etendue, as the product of the effective area $dS\cos(\theta)$ projected perpendicular to the direction of propagation and the solid angle $d\Omega$. The $A\Omega$ multiplies the brightness: the greater the detector area and/or the solid angle Ω subtended by the source, the greater the collected power (Figure 2.3):

$$dA\Omega = dS \cos \theta \, d\Omega. \tag{2.34}$$

Given a surface $d\Sigma$ belonging to a source Σ, and a surface dS belonging to a receiving system S, we can define a ray of light, a beam of cylindrical flux coming from an infinitesimal section $d\Sigma$ and incident on dS at the distance d from the source. It will subtend a solid angle $d\Omega_S$ or $d\Omega_\Sigma$, if seen from the surface Σ or S (Figure 2.4).

On one hand, we have that:

$$A_{\Sigma S} = d\Sigma \cos \theta_\Sigma \tag{2.35}$$

$$\Omega_{\Sigma S} = dS \cos \theta_S / d^2 \tag{2.36}$$

$$A\Omega_{\Sigma S} = d\Sigma \cos \theta_\Sigma dS \cos \theta_S / d^2. \tag{2.37}$$

On the other hand, we have that:

$$A_{S\Sigma} = dS \cos \theta_S \tag{2.38}$$

$$\Omega_{S\Sigma} = d\Sigma \cos \theta_\Sigma / d^2 \tag{2.39}$$

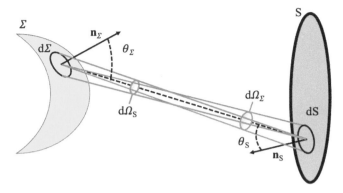

Figure 2.4. $A\Omega$ conservation.

$$A\Omega_{S\Sigma} = dS \cos \theta_S d\Sigma \cos \theta_\Sigma / d^2. \tag{2.40}$$

It is thus evident that the throughput is conserved in this simple system:

$$A\Omega_{S\Sigma} = A\Omega_{\Sigma S}. \tag{2.41}$$

So, the throughput of a light beam is preserved as the radiative power along the beam. From a source's point of view:

$$A_{\text{source}}\Omega_{\text{pupil}}. \tag{2.42}$$

From the detector's point of view (and the same is true within a refractive or reflective optical system):

$$A_{\text{pupil}}\Omega_{\text{source}}. \tag{2.43}$$

In the propagation of the power along a light beam, we have, for the brightness:

$$B_\Sigma = \frac{\Delta W}{A\Omega_{\Sigma S}} = \frac{\Delta W}{A\Omega_{S\Sigma}} = B_S. \tag{2.44}$$

So, the brightness is also conserved along the light beam.

2.3 Geometrical Optics: Snell's Law

When an electromagnetic wave goes from one medium to another, due to the refractive index, its velocity changes from:

$$vel_1 = c/n_1 \tag{2.45}$$

to

$$vel_2 = c/n_2 \tag{2.46}$$

where c is the speed of light in a vacuum, and n_1 and n_2 are the refractive indexes of the two media 1 and 2.

In fact, the frequency of the wave remains the same (otherwise the wave, at the interface, would be discontinuous), and so the electromagnetic-wave velocity is the quantity that changes when going from medium 1 to medium 2. Thus, we have that the frequencies in the two media are related as in the following:

$$\nu_1 = \nu_2 = \nu. \tag{2.47}$$

Consequently, the wavelengths in the two media differ as in the following:

$$\lambda_1 = \frac{vel_1}{\nu} \tag{2.48}$$

and

$$\lambda_2 = \frac{vel_2}{\nu} \tag{2.49}$$

thus

$$\lambda_1 = \frac{vel_1}{\nu} = \frac{c}{n_1\nu} \rightarrow \nu = \frac{c}{n_1\lambda_1} = \frac{c}{n_2\lambda_2} \tag{2.50}$$

and so:

$$n_1\lambda_1 = n_2\lambda_2. \tag{2.51}$$

This equation relates the wavelength of an electromagnetic wave and the refractive index of a medium. From this, we can derive a proper definition of the refractive index of a medium as the ratio between the wavelengths in the vacuum and in the medium itself.

Suppose we have a wave that hits a point **O** of the separation surface, the interface, between two media of refractive indices n_1 and n_2 with propagation direction θ_1 with respect to the normal along the surface itself (Figure 2.5).

The incident wave has an electric field (k = wavevector, which describes its propagation) E_i:

$$E_i = E_{0i}e^{[i(k_i r - \omega t)]}. \tag{2.52}$$

We will then have a reflected wave E'_i and a refracted wave E_r, so, their electric fields will be:

$$E'_i = E'_{0i}e^{[i(k'_i \cdot r - \omega t)]} \tag{2.53}$$

$$E_r = E_{0r}e^{[i(k_r r - \omega t)]}. \tag{2.54}$$

Since, in order to have continuity at the interface, the fields must be equal, and considering the projection along the interface plane (we call the x-axis the one parallel to the line drawn) of the wavevector, and remembering that $k = 2\pi/\lambda$, we have that:

$$n_1\lambda_1 = n_2\lambda_2. \tag{2.55}$$

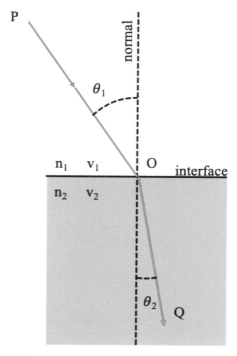

Figure 2.5. Refraction of light through an interface between two media of refractive indices n_1 and n_2.

As a consequence:

$$k_{xi}x = k'_{xi}x = k_{xr}x \rightarrow \frac{2\pi}{\lambda_i}\sin\theta_i = \frac{2\pi}{\lambda'_i}\sin\theta'_i = \frac{2\pi}{\lambda_r}\sin\theta_r. \qquad (2.56)$$

And given the relationship between the wavelengths, we have that $\theta'_i = \theta_i$ (specular reflection), while:

$$\lambda_i = \frac{n_2}{n_1}\lambda_r \rightarrow \frac{2\pi}{\lambda_r n_2/n_1}\sin\theta_i = \frac{2\pi}{\lambda_r}\sin\theta_r \rightarrow n_1\sin\theta_i = n_2\sin\theta_r. \qquad (2.57)$$

Equation (2.57) is called Snell's law and relates the refractive index of a medium with the angle, with respect to the normal, of the propagation of the wave in the two media. It is clear that when an electromagnetic wave goes from a medium with a lower refractive index to one with a higher index, the radiation approaches the normal to the interface.

The refractive index of a medium is not a constant but depends on the wavelength (Figure 2.6). Snell's law can thus be rewritten as:

$$n_2(\lambda_2)\sin\theta_r = n_1(\lambda_1)\sin\theta_i. \qquad (2.58)$$

In the case of a polychromatic wave propagating in a medium with two parallel faces, radiation will be deflected at the input surface with different angles depending on the wavelength, but will then recombine at the output surface into the same polychromatic wave. If, instead, we consider an optical element that does not have parallel faces (a prism), it can decompose the light when it enters the medium and keep on decomposing it at the output of the optical element. Under this principle is the rainbow principle: an incoming beam of sunlight, after being refracted by a drop of water in the atmosphere, is reflected on the back of a drop of water back and then refracted again; the violet light ($\lambda \sim 400$ nm) is refracted more than the red light ($\lambda \sim 650$ nm); (Figure 2.7).

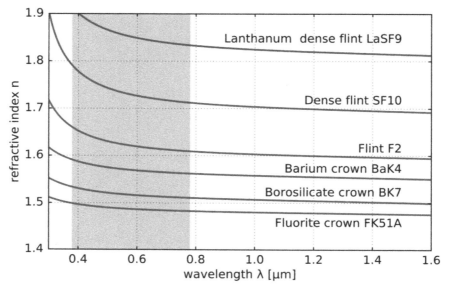

Figure 2.6. Dependence of the refractive index on wavelength for various glasses. The shaded zone indicates the range of visible light. Credit: Wikimedia: (CC BY-SA 4.0) https://commons.wikimedia.org/wiki/File: Prism_rainbow_schema.png.

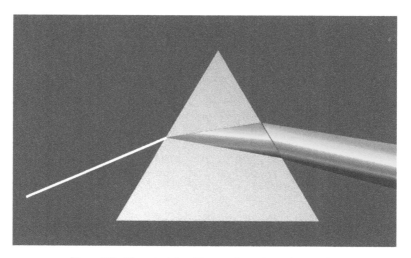

Figure 2.7. The principle of how a dispersion prism works.

2.4 Lenses and Mirrors

2.4.1 Lenses

In geometric optics we neglect all of the mechanisms of physical optics such as diffraction, and we assume that the light is formed by light rays that travel in a straight line. From a geometric point of view, it is customary to define *objects* whose emission is to be revealed in an *image*: *object* and *image* are conjugate points of an optical system. *Objects* and *images* can be real or virtual: in the first case the *object* or *image* is the center of the rays, while in the second case *object* or *image* is on the extension of these rays, where they would meet.

The *focus* of an optical system is the point conjugated to an infinity point. We can define the following conventions:

- *V* is the vertex, usually the center of an optical element.
- Position *p* is positive if it falls to the left of *V*.
- Position *q* is positive if it falls to the right of *V*.
- Curvature radius *R* is positive if the center of the sphere falls to the right of *V*.

Depending on the geometry of the surface, we can create refocused images, disperse them, enlarge them, make them smaller, etc. Most lenses have flat and/or spherical surfaces (concave or convex) with curvature radii that determine their properties (see Figure 2.8). A biconvex or plane-convex lens is convergent: rays coming from infinity focus at a point at a distance *f* from the lens after the same. A biconcave or concave plane lens is called divergent: the focus is on the source side. Convex-concave lenses can be divergent or converging. According to Snell's law, incident rays on the surface of a lens with $n > 1$ approach normal.

Given a thin lens, such as the one in Figure 2.9, three rays determine the characteristics of the *image* and of the *object* (see Figure 2.10):

- A ray, starting from the *object*, parallel to the lens axis, that is directed into the focal point by the lens
- A ray passing through the lens center, which goes through the lens unaffected

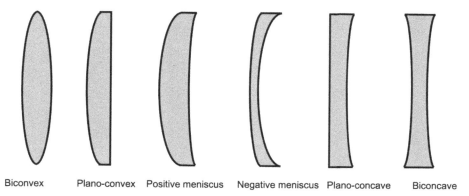

Biconvex   Plano-convex   Positive meniscus   Negative meniscus   Plano-concave   Biconcave

Figure 2.8. Types of lenses.

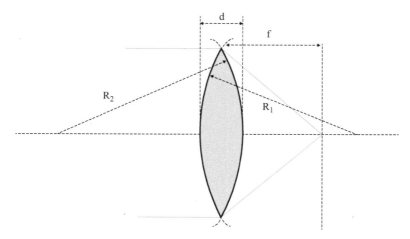

Figure 2.9. A convergent lens. The focal point is the conjugate of a point to infinity.

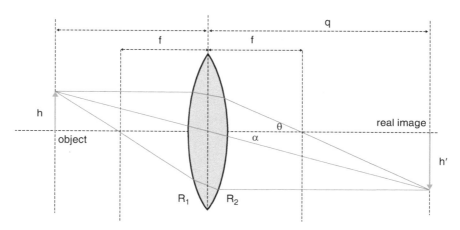

Figure 2.10. Relation between conjugate points of a thin lens.

- A ray starting from the object, passing through the focus, on the left side of the lens, which is than deflected into an axis parallel to the lens axis beyond the lens.

Given these three rays, we can define the lens parameters as in the following. The magnification M is the ratio between the dimension of the image h' and the dimension of the object h. We have:

$$M = \frac{h'}{h} = \frac{-q \cdot \tan \alpha}{p \cdot \tan \alpha}. \tag{2.59}$$

However,

$$\tan \theta = \frac{h}{f} = -\frac{h'}{q-f} \rightarrow \frac{h'}{h} = -\frac{q-f}{f} = -\frac{q}{p} \rightarrow = \frac{1}{p} + \frac{1}{q} = \frac{1}{f}. \tag{2.60}$$

The latter is the equation of a thin lens for the conjugate points at the distances p and q and how they are linked with the focal point at the distance f from the lens vertex (Figure 2.10).

In addition, the focal point at the distance f is linked to the radii of curvature R_1 and R_2 of the lens as in the following:

$$\frac{1}{f} = (n - 1)\left(\frac{1}{R_1} + \frac{1}{R_2}\right). \tag{2.61}$$

2.4.2 Mirrors

A flat mirror does not produce magnification, but instead allows one to "bend" an optical beam, bringing an image toward an eye or a detector, for example (Figure 2.11). If θ_i is the incident angle with respect to normal and θ_r is the reflected angle, we have that P is the position of an object and Q is the position of the image. We have:

$$\sin \theta_i = \sin \theta_r. \tag{2.62}$$

A concave, spherical mirror with a radius of curvature r can be treated in a similar way to a thin lens (Figure 2.12). A ray passing through the center C will be reflected back with the same angle, while a ray hitting the mirror at its vertex V will be reflected as if it were a flat mirror. A spherical mirror produces a magnification M, a mirroring of the image, and a relationship between the object at position O

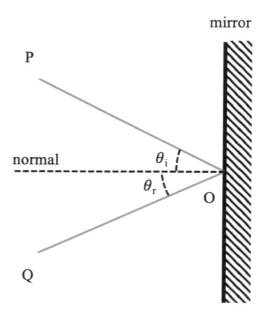

Figure 2.11. Reflection onto a flat mirror.

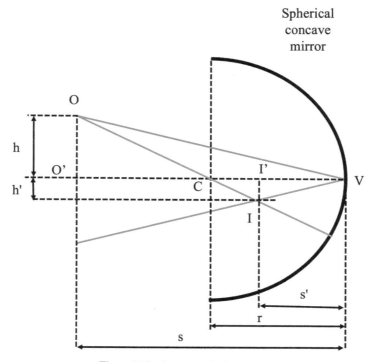

Figure 2.12. Concave spherical mirror.

and distant s from the mirror, and an image at position I and distant s' from it (see Figure 2.12), assuming α to be the angle of the rays at the vertex, as in the following:

$$M = \frac{h'}{h} = \frac{-s' \cdot \tan \alpha}{s \cdot \tan \alpha} = -\frac{s'}{s}. \tag{2.63}$$

That said, we have:

$$\tan \alpha = \frac{h}{s - R} = -\frac{h'}{R - s'} \rightarrow \frac{h'}{h} = -\frac{R - s'}{s - R} = -\frac{s'}{s} \rightarrow \frac{2}{R} = \frac{1}{s} + \frac{1}{s'} = \frac{1}{f}. \tag{2.64}$$

A spherical mirror has a focal length of $R/2$. We remind the reader that the focal length is the conjugate point of an image at infinity. In this sense, however, a paraboloid mirror offers a better performance than does a spherical mirror.

2.4.3 Lens $A\Omega$

If we now think of the power W emitted by an extended source of brightness B and detected by a detector, its power should be integrated on the whole area A_d of the

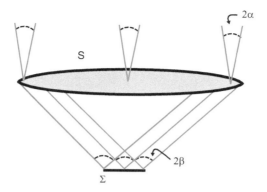

Figure 2.13. $A\Omega$ conservation.

detector and on the whole solid angle ω_s within the source, to determine the total power. We thus have:

$$W = \int_{A_d} \int_{\omega_s} B dA_d \cos\theta \, d\omega_s. \tag{2.65}$$

For example, if the source is distant, we can separate the integrals:

$$W = B \int_{A_d} dA_d \int_{\omega_s} \cos\theta \, d\omega_s = BA_d\Omega_s = BA_s\Omega_d \tag{2.66}$$

where we used the conservation of the throughput for which the source area A_s times the detector solid angle Ω_d is the same as the detector area A_d times the source solid angle Ω_s (Figure 2.13).

Imagine we want to calculate the $A\Omega$ of an optical system like a convergent lens. Let the convergent lens be of diameter S and a detector be of diameter Σ (see Figure 2.13).

We have:

$$A_{in}\Omega_{in} = \pi\frac{S^2}{4} \cdot \pi\alpha^2 \tag{2.67}$$

and

$$A_{out}\Omega_{out} = \pi\frac{\Sigma^2}{4} \cdot \pi\beta^2. \tag{2.68}$$

For a thin lens,

$$\frac{S}{2f} = \tan\beta \simeq \beta \rightarrow \frac{1}{f} = \frac{2\beta}{S}. \tag{2.69}$$

From the definition of focal distance, considering the central radius:

$$\frac{\Sigma}{f} \simeq 2\alpha \rightarrow \frac{1}{f} = \frac{2\alpha}{\Sigma} \rightarrow \frac{\alpha}{\Sigma} = \frac{\beta}{S} \rightarrow \Sigma\beta = S\alpha. \tag{2.70}$$

So, we have that:

$$A_{\text{in}}\Omega_{\text{in}} = \pi\frac{S^2}{4} \cdot \pi\alpha^2 = \pi\frac{\Sigma^2}{4} \cdot \pi\beta^2 = A_{\text{out}}\Omega_{\text{out}} \tag{2.71}$$

which demonstrates the conservation of $A\Omega$ for a thin lens and can be extended to any optical system.

2.5 Diffraction and Interference

For some applications, it is necessary to consider light as a wave and not as a beam of particles. This is necessary when the wavelengths are of dimensions comparable to the dimensions of the optical systems. Huygens' principle states the following: *given a source of electromagnetic waves, at every moment of wave propagation, each point of the wave front can be considered as a source of spherical waves. The new wave front will be composed from the envelope of the front's elementary wave.*

This is a useful principle to explain the behavior of radiation when an obstacle obscures it and its capability to go around the obstacles themselves. When the obstacle in question is a hole or a slit, the wave front changes appearance: the incident front is flat and the emerging front is a flat front only in the central area (Figure 2.14).

If the slit is reduced to zero, the new front is spherical. The Huygens principle allows us to understand the capacity of electromagnetic waves to get around the obstacles. It also allows us to understand the interference and diffraction figures of single and multiple slits (a grating).

Consider a slit as in Figure 2.15, with a flat wave front that hits it.

Suppose we put a screen S at infinity (or use a lens to simulate infinity). Let x be the coordinate of any point of the slit, the origin of the new wave front. What happens at the points P of the screen when all of the elementary sources on the slit are combined as a function of θ? In order to consider the contribution of the elementary sources of the new wave front, we will have to integrate all of the contributions between 0 and d. We assume the wave is in a cosine-like form. Consider the effect of the phase delay of $x\sin(\theta)$.

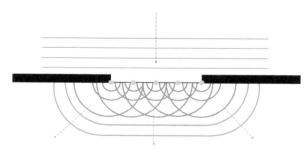

Figure 2.14. Electromagnetic wave arising from a slit. The Huygens principle predicts that the electromagnetic wave, beyond the slit, can be viewed as the superposition of the electromagnetic waves generated at the slit.

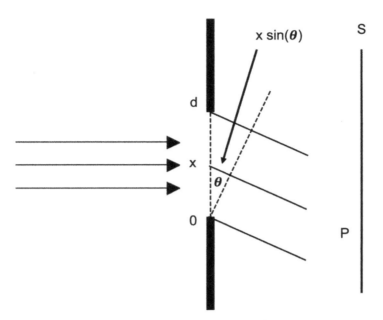

Figure 2.15. Slit hit by an electromagnetic wave.

The resulting field will be:

$$E(P, t) = a \int_0^d \cos[kx \sin(\theta) - \omega t] dx. \tag{2.72}$$

Since $\cos(\alpha - \beta) = \cos(\alpha)\cos(\beta) + \sin(\alpha)\sin(\beta)$,
we have that:

$$E(P, t) = a \int_0^d \cos[kx \sin(\theta)]\cos(\omega t) + \sin[kx \sin(\theta)]\sin(\omega t) dx. \tag{2.73}$$

Changing a variable,

$$y = kx \sin \theta \rightarrow \frac{dy}{k \sin \theta} = dx \tag{2.74}$$

we get:

$$E(P, t) = \frac{a}{k \sin(\theta)}[A \cos(\omega t) + B \sin(\omega t)] \tag{2.75}$$

where

$$A = \sin(kd \sin(\theta))$$

$$B = 1 - \cos(kd \sin(\theta)).$$

To calculate the field intensity $I(P)$, we have to calculate the mean value of the square of the electric field E: $<E^2>$.

Note that:

$$\langle \cos^2(\omega t) \rangle = \langle \sin^2(\omega t) \rangle = \frac{1}{2}; \langle \sin(\omega t) \cos(\omega t) \rangle = 0.$$

We get:

$$I(P) = \langle E^2(P, t) \rangle = \frac{a^2}{2[k \sin(\theta)]^2}(A^2 + B^2) \tag{2.76}$$

and, therefore,

$$I(P) = \frac{a^2}{2[k \sin(\theta)]^2}(\sin^2[kd \sin(\theta)] + 1 + \cos^2[kd \sin(\theta)] - 2$$
$$\cos[kd \sin(\theta)]). \tag{2.77}$$

From which we get:

$$I(P) = \frac{a^2}{[k \sin(\theta)]^2}(1 - \cos[kd \sin(\theta)]) = \frac{d^2}{d^2}\frac{a^2}{[k \sin(\theta)]^2}2\sin^2\left[\frac{kd \sin(\theta)}{2}\right]. \tag{2.78}$$

Now, we put:

$$\alpha = \frac{kd \sin(\theta)}{2} = \frac{\pi d}{\lambda}\sin(\theta)$$

from which we obtain:

$$I(P) = \frac{a^2 d^2}{2}\left[\frac{\sin^2(\alpha)}{\alpha^2}\right]. \tag{2.79}$$

We get the classic interference figure with:
- Light bands for semi-integers α, $\alpha = \pm\frac{2n + 1}{2}\pi$
- Dark bands for integer α, $\alpha = \pm n\pi$.

This occurs for $\sin(\theta)$ multiples of λ/d. The central peak, seen from the slit, has an angular width of:

$$2 \sin(\theta) \cong 2\theta = 2\frac{\lambda}{d}$$

which will be greater given a greater λ and a smaller d.

Taking the rays that come from infinity (a flat wave) and focusing them with a lens can be seen as building a telescope that can be approximated as an opening slit d. By observing a celestial object, we will obtain a diffraction figure such as the one shown in Figure 2.16. If we look at two neighboring objects at the same time, we will get two similar diffraction figures, and if they are too close, one cannot distinguish them. Rayleigh's criterion tells us that two objects can be distinguished (the ability to distinguish two object is called angular resolution) if the maximum of the diffraction pattern of the first is further than the first zero of the other diffraction figure.

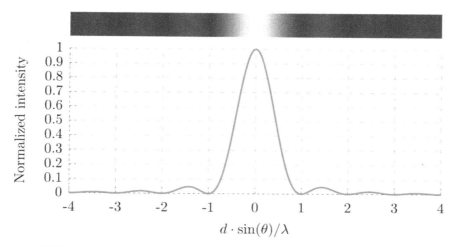

Figure 2.16. Diffraction pattern from a single slit. We have maxima at semi-integers $d \sin(\theta)/\lambda$ and minima at integers $d \sin(\theta)/\lambda$. Credit: Wikipedia: DL6ER (CC BY-SA 4.0).

$$\Delta\theta_{\min}^{\text{slit}} = \frac{\lambda}{d} \tag{2.80}$$

If instead of a slit we had a circular opening, we must correct that number by a factor of 1.22.

$$\Delta\theta_{\min}^{\text{circ}} = 1.22\frac{\lambda}{d} \tag{2.81}$$

which is the diffraction limit, expressed in radians, for the angular resolution of a telescope of diameter d, working at wavelength λ.

Let's now have an aperture telescope D whose resolution is dictated by diffraction. We will have a throughput $A\Omega$ from a measure of the area A:

$$A = \pi\frac{D^2}{4} \tag{2.82}$$

and of the solid angle Ω:

$$\Omega = \pi\frac{(\Delta\theta_{\min}^{\text{circ}})^2}{4} = \pi\frac{\left(1.22\frac{\lambda}{D}\right)^2}{4}. \tag{2.83}$$

Thus, the throughput:

$$A \cdot \Omega = \pi\frac{D^2}{4} \cdot \pi\frac{\left(1.22\frac{\lambda}{D}\right)^2}{4} \cong \lambda^2.$$

Figure 2.17. The Chinese Five-hundred-meter Aperture Spherical Telescope, FAST. Credit: Wikipedia: Absolute Cosmos (CC BY 3.0).

So, in the diffraction limit, in order to increase the angular resolution, we have to increase the telescope radius. For this reason, large radio telescopes are built, such as the giant Chinese Five-hundred-meter Aperture Spherical Telescope, FAST (see Figure 2.17). However, the power hitting a detector will not increase, as it will be proportional to λ^2.

Experimental Astrophysics

Elia Stefano Battistelli

Chapter 3

Telescopes

Basic working principles of telescopes are given in this chapter. Thin lenses, mirrors, and the conservation of the throughput in an optical system are described. Different kinds of telescopes and mounts including their use, construction, advantages, and disadvantages are presented. Refractive telescopes, reflective telescopes, in-axis and off-axis telescopes, and single and multimirror telescopes are presented, as well as some highlights on aberrations. The concepts of the focal plane, size on the focal plane, field of view, and angular resolution are presented as well.

3.1 Diffraction-limited Telescopes or Not?

The simplest telescope one can think of is a lens with a focal length. Objects along the optical axis produce a beam parallel to the optical axis itself and are focused at the focal point (Figure 3.1). Objects that send their radiation toward the telescope in a direction other than the optical axis, at first approximation, focus on a point adjacent to the focus at the same distance as the lens. We can thus define a focal plane that can be used to place a camera to record the brightness distribution of the sky: angles in the sky translate to spatial distances on the focal plane and images can indeed be made.

We remind the reader of the conservation of throughput, which tells us that observations on large fields of view require large focal planes and large cameras coupled to our telescopes. Taking the rays that come from infinity (a plane wave) and focusing them with a lens means building a telescope that can be approximated as an opening slit d or a circular aperture. So, by observing a celestial object at the wavelength λ, we will obtain a diffraction figure that limits our angular resolution to 1.22 λ/d radians. In addition, an aperture telescope D whose field of view is dictated by diffraction will have a throughput $A\Omega \sim \lambda^2$. So, as already stated, when the optics of a telescope is limited by diffraction, increasing the diameter of the telescope improves the angular resolution, but does not increase the incident power.

Using a telescope that is diffraction limited is a choice that has consequences in terms of angular resolution and the power hitting the detectors. Also, whether to use

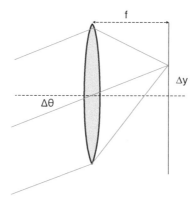

Figure 3.1. Focal plane tilt of Δy when a source in the sky is not observed on the optical axis of a telescope or lens.

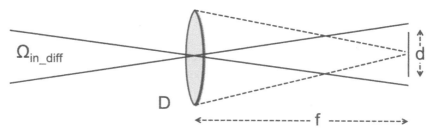

Figure 3.2. Non optimized diffraction-limited system.

a telescope that is diffraction limited or not depends on the observational conditions and scientific requirements. These characteristics depend in turn on the focal length f of the telescope. We can have three different kind of systems:

- Diffraction-limited system not optimized: the incoming diffraction-limit angle, Ω_{in_diff}, from the sky spreads onto an area on the focal plane larger than the pixel size d. This solution is used to oversample a detector array (Figure 3.2).
- Optimized diffraction-limited system: the incoming diffraction-limit angle, Ω_{in_diff}, from the sky spreads onto an area that has the dimensions of the detector, d. This solution is chosen, mainly at radio frequencies, to increase the angular resolution of a telescope to the maximum (Figure 3.3).
- Non diffraction-limited system: the incoming diffraction-limit angle, Ω_{in_diff}, from the sky spreads onto an area that has dimensions smaller than the detector size (Figure 3.4). Thus, on the actual detector, radiation from a larger solid angle is focused. This means that the telescope has an angular resolution that is not as high as the diffraction-limited one. However, the throughput (with the same telescope size) is larger. This solution is adopted when a telescope needs to be more luminous, or the angular resolution is

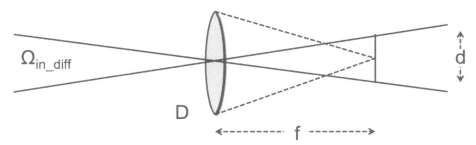

Figure 3.3. Optimized diffraction-limited system.

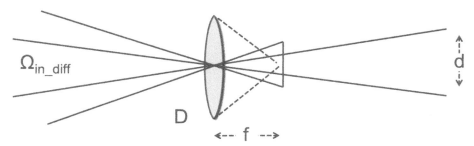

Figure 3.4. Non diffraction-limited system.

limited by other factors such as the atmosphere, and so it is useless to reach the diffraction limit.

As mentioned, one of the reasons why the angular resolution may not be that dictated by diffraction is the Earth's atmosphere. An undisturbed wave front is disturbed by vortices present in the atmosphere. The parameter r_0 is an estimate of the coherence of the wave front. "Seeing" generates three effects (studied empirically):

- Enlargement: angular resolution is no longer dominated by diffraction, but by r_0:

$$1.22\frac{\lambda}{D} \rightarrow 1.22\frac{\lambda}{r_0}. \tag{3.1}$$

- The source moves with a variance that is also determined by r_0 and by the diameter D:

$$\sigma_x^2 = \sigma_y^2 = 0.18 \cdot D^{1/3}r_0^{-5/3}. \tag{3.2}$$

- Scintillation: the source changes intensity. The relative variance of its brightness will depend on the wavelength λ and on the zenith angle z:

$$\frac{\sigma_I^2}{I^2} = 19.12 \cdot \lambda^{-\frac{7}{6}} \quad \cos(z)^{-11/6}. \tag{3.3}$$

3.2 Refractive Telescopes

A telescope is a tool for collecting electromagnetic waves. As mentioned, fundamental elements are the size of the lens (primary mirror) and the focal length. The $f\#$ is the ratio between the focal length and the diameter of the lens (mirror or lens):

$$f\# = \frac{f}{D}. \tag{3.4}$$

Telescopes can be "fast" or "slow" depending on how large the focus and the lens or primary mirror diameter is. A "fast" telescope converges to the focus rapidly, so the $f\#$ is small. A "slow" telescope is one with a large $f\#$. The solid angle with which a pixel of a receiver sees the primary telescope element is:

$$\alpha \cong \frac{D}{2}\frac{1}{f} \rightarrow \Omega \cong \pi\alpha^2 \cong \frac{\pi}{4}\frac{1}{(f/D)^2} = \frac{\pi}{4}\frac{1}{(f\#)^2}. \tag{3.5}$$

So, in order to increase the power that hits a pixel, we need to increase Ω and then decrease the $f\#$ to make it more luminous.

Refractive telescopes, in optical astronomy, have mainly historical or amateur importance. In millimeter astronomy, they are still used. In general, they are made up of a lens (an objective) and an ocular (an eyepiece) with the distance linked to their focal lengths. The objective bends parallel rays into a focus at distance f_1. The eyepiece allows one to focus the light into the observer's eye. The magnification is the relationship between the field of view of the objective and the eyepiece. For an infinite-distance source (which holds for the astronomical case), it is equal to the ratio of the focal lengths of the objective and the eyepiece. It can clearly be modified by changing the eyepiece:

$$M = \frac{f_1}{f_2}. \tag{3.6}$$

Several configurations and combination of different kinds of lenses exist for specific cases and to correct for defects in the image such as aberrations that cause light coming from one direction on the sky to be spread over a finite-sized zone on the focal plane as opposed to a single point.

Galileo telescope: Of fundamental importance is the Galileo telescope that Galileo Galilei built in 1609. This is the first telescope ever built, although the first lenses had already been built a few years before by Dutch opticians (Figure 3.5). Its importance is connected to the following points:

- It uses two lenses: one plane-convex or biconvex (converging) lens and one plane-concave or biconcave (divergent) lens.
- The lenses are placed in such a way that the second focus of the objective coincides with the first focus of the eyepiece.
- The intermediate focus (for the ocular) is imaginary (it is located on the opposite side with respect to the eye position), so it does not invert the image.
- It allowed Galileo Galilei to observe the craters on the moon, the moons of Jupiter, and the phases of Venus (Figure 3.6).

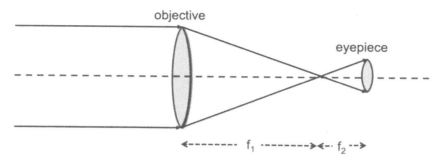

Figure 3.5. Two-lens telescope.

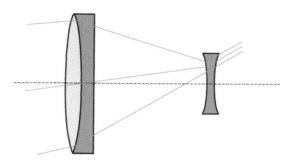

Figure 3.6. The Galileo telescope.

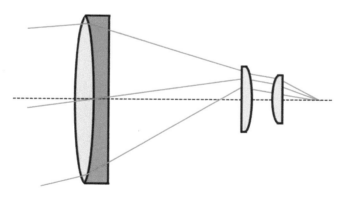

Figure 3.7. The Kepler telescope.

Kepler telescope: The Kepler telescope, as well as the Fraunhofer achromatic refractors (1818), uses a convergent (biconvex) lens that allows for the increase of the field of view. However, the image is reversed (Figure 3.7).

The largest refracting telescope in the world (~1.02 m) is that of the Yerkes Observatory (Chicago, 1893; Figure 3.8).

As stated, the focal plane is the plane at the focus distance from the objective. The focal plane scale (typically arcseconds per millimeter) links the linear dimensions of

Figure 3.8. The Yerkes 40 inch telescope backdrops Albert Einstein's visit in 1921 May. Credit: Yerkes Observatory (Public Domain).

the focal plane to the angular dimensions in the sky. For a telescope with $f = 2$ m, the focal plane scale is $s = 1/f \sim 100''$ mm^{-1}.

As mentioned already, the simplest telescope is a converging lens with focal length f. Given an object (with a brightness distribution $B(\alpha, \delta)$) placed at infinity, the telescope produces an image of it on the focal plane where the detector is placed. If the detector is composed of multiple pixels, each of area A_p, each of them will collect the power coming from a solid angle in the sky Ω_{in}. Given the lens area A_L and the exit solid angle Ω_{out},

$$\Omega_{\text{in}} A_L = \Omega_{\text{out}} A_p \rightarrow \Omega_{\text{in}} = \Omega_{\text{out}} \frac{A_p}{A_L} \rightarrow \Omega_{\text{in}} = \frac{A_L}{f^2} \frac{A_p}{A_L} = \Omega_{\text{in}} = \frac{A_p}{f^2}. \tag{3.7}$$

And so, given a brightness in the sky $B(\alpha, \delta)$, the collected power by each pixel is therefore:

$$W = A_L \Omega_{\text{in}} B(\alpha, \delta). \tag{3.8}$$

We therefore have a trade-off problem between high angular resolutions in the sky (i.e., small Ω_{in}) and high sensitivity (i.e., high W). For the same angular resolution, it is therefore convenient to increase the dimensions of the primary radiation collector (i.e., A_L) and thus have large telescopes. In terms of $f\#$, the previous equation can be rewritten as:

$$W = \pi \left(\frac{D}{2} \right)^2 \frac{A_p}{f^2} B(\alpha, \delta) = \frac{\pi}{4} \frac{A_p}{f\#^2} B(\alpha, \delta). \tag{3.9}$$

So, slow telescopes (large $f\#$) are not very luminous but have a high angular resolution, while fast telescopes (small $f\#$) have a low angular resolution but high brightness.

Unfortunately, lenses made of dielectric materials (glass, polyethylene, etc.) have a refractive index that depends on the wavelength λ, and therefore chromatic (aberration) effects will be present. This can be partially corrected for, but for large telescopes, it is preferred to use mirrors.

3.3 Reflective Telescopes

Reflective telescopes, unlike refracting telescopes, have the advantage that radiation does not pass through optical elements other than air, and, therefore, in principle, have achromatic performances. To be totally fair, mirrors are made of metals, and unfortunately also for metals, the reflectivity is a function of λ (Figure 3.9). However, in most cases this effect is smaller and can be neglected. Aluminum (Al) is the most versatile material that maintains a high reflectivity from the UV to the IR. Silver (Ag), and to some extent gold (Au), however, have greater reflectivity in the optical–IR band, but lower performances in the UV. Usually, silver or aluminum are deposited onto "optical" glass (a low thermal expansion material).

Neglecting chromatic effects, we can consider conical surfaces (paraboloids, ellipsoids, hyperboloids, and spheres) that convey optical rays from one focus to another and have a cylindrical symmetry around the optical axis. In cylindrical coordinates (ρ, φ, z), a conic is symmetric in φ, which therefore does not appear in the following equation:

$$\rho^2 - 2Rz + (1 - e^2)z^2 = 0 \qquad (3.10)$$

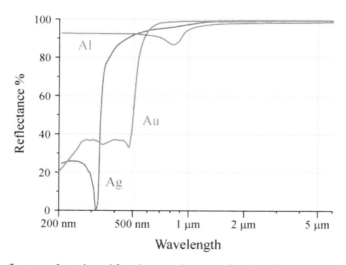

Figure 3.9. The reflectance of metals used for telescope mirrors as a function of wavelength. Credit: Wikipedia: DrBob (CC BY-SA 3.0).

where R is the curvature radius measured in the vertex and e is the eccentricity. For a single mirror telescope, it is evident that the paraboloid is the most suitable geometry, as it has a focus at infinity (in the sky) and a second one on the focal plane. Sometimes, for ease of construction, the parabolic mirror is replaced by a spherical mirror, and, in this case, spherical aberrations occur.

Newtonian telescope: The Newtonian telescope (1670) consists of a parabolic primary (objective) mirror and a 45° flat folding mirror with an ellipsoidal section. It does not have a long focal length. Its simplicity has made it very popular for different kinds of applications, including amateur telescopes. It can be characterized by a large field of view.

The field of view of a telescope is dictated by deviations from ideality. Optical aberrations are a characteristic of an optical system that focuses the light onto a point, and cause the light to be spread over a finite-sized region. Several reasons can cause such aberrations and are thus divided in different classes:

- Chromatic aberration: lenses with a refractive index that depends on the wavelength and mirrors for which reflectivity also depends on the wavelength have different optical properties depending on the wavelength itself. Achromatic doublets, made of two different glasses, a crown glass, with low refractive index, and a flint glass, with high refractive index, can compensate for this deviation (Figure 3.10).
- Spherical aberration: mirrors of shapes other than a paraboloid concentrate radiation in an area of the focal plane of finite size. Schmidt telescopes mitigate this issue using a correcting lens in front of it (Figure 3.11).
- Off-axis aberrations: a paraboloid works perfectly only if radiation arrives along the optical axis. Within a field of view, radiation comes also from the off-axis direction and deviations from ideality arise. By developing in series the light rays' distribution, we get three terms that produce different effects:
  - Astigmatism: a third-order term, astigmatism is symmetrical with respect to the divergence from the optical axis, which makes the focusing of meridional and sagittal planes to converge at different positions (Figure 3.12).

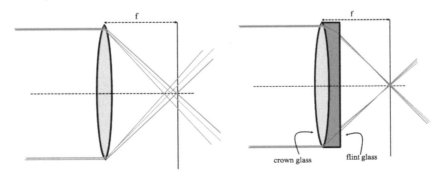

Figure 3.10. Left: chromatic aberration. Right: correction of it using an achromatic doublet.

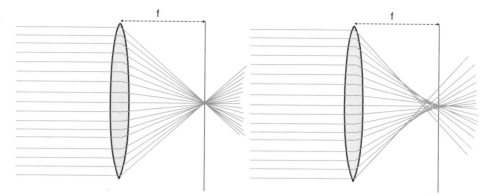

Figure 3.11. Left: lens made with a parabolic surface with no spherical aberration. Right: lens made with a spherical surface where a spherical aberration can be seen.

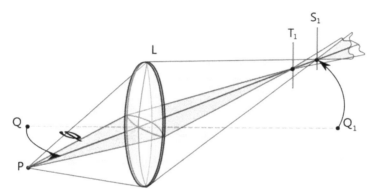

Figure 3.12. Astigmatism: the meridional and sagittal planes focus in different positions. Credit: Wikipedia: Sebastian Kroch (CC BY-SA 3.0).

- ○ Coma: a second-order term, coma creates an asymmetric image. This aberration makes an image on the sky distorted in an asymmetric way with respect to the incidence angle. This effect is mitigated in spherical mirrors where there is no preferential direction. Different telescopes mitigate this issue, such as the Ritchey–Chrétien configuration (Figure 3.13). One way of mitigating the effect is to use a Barlow lens, a divergent lens that increases the effective focal length of a telescope.
- ○ Distortion: this is a deformation of the image without a real degradation of it. The net effect is a variation of the magnification that varies as a function of the distance from the center (Figure 3.14).
- • Field curvature: this aberration, also known as Petzval field curvature, is due to the fact that the focal plane is distributed on a curved plane, and so it cannot be brought properly into focus on a flat surface. One way to reduce this effect is to insert an aperture stop in order to remove edge light rays.

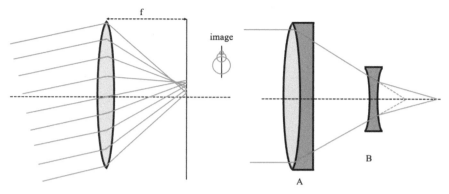

Figure 3.13. Left: coma aberration for off-axis rays. Right: a Barlow lens is placed at position B and increases the focal length of the telescope.

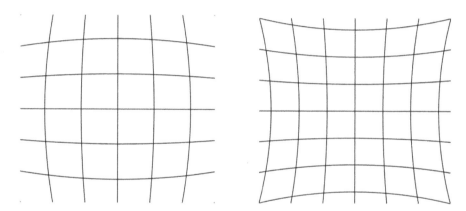

Figure 3.14. Distortion: deformation of the images with increasing distance from the optical axis.

This method, however, reduces the light collecting power of the lens. Another way is, clearly, to provide a curved focal plane (Figure 3.15).

A system of two (or more) mirrors allows the optical system to be more compact and partially compensates for aberrations. This has advantages not only for the construction of the telescope, but also for using it more efficiently; for example, by better counterbalancing any imbalances.

Cassegrain telescope: In a Cassegrain (1672) telescope, the focus of the primary mirror (a paraboloid) is made to coincide with one of the foci of a secondary mirror (a hyperboloid). The focal plane is on the other focus behind the primary. This configuration allows the focal length to be very long (Figure 3.16).

Gregorian telescope: The Gregorian telescope (1663) combines a parabolic primary mirror with an elliptical secondary mirror. It is a configuration similar to the Cassegrain (Figure 3.17).

In general, the same focal length (and the same $f\#$) can be obtained by simultaneously changing the conical constants of the mirrors. These quantities

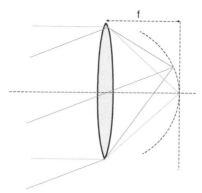

Figure 3.15. Field curvature: this aberration arises from the curved focal plane imaged on a flat plane.

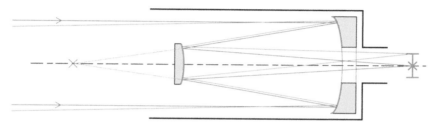

Figure 3.16. A Cassegrain telescope. Credit: Wikipedia: Krishnavedala (CC BY-SA 4.0).

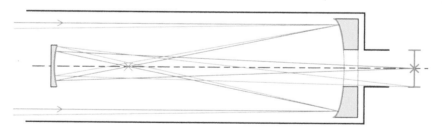

Figure 3.17. A Gregorian telescope. Credit: Wikipedia: Krishnavedala (CC BY SA 4.0).

may vary slightly by minimizing spherical aberrations and coma by compensating for any distortions with other distortions. Usually a numerical ray tracing is solved, or the equations are solved to zero out the coma and other aberrations.

Ritchey–Chrétien telescope: The Ritchey–Chrétien telescope (R–C, 1910) has two hyperbolic mirrors. It eliminates coma and spherical aberrations, but has problems with astigmatism and curvature of the field. Most large telescopes are R–C (e.g., the largest optical telescope, the Gran Telescopio Canarias, GTC)[1]. The design

---

[1] http://www.gtc.iac.es/

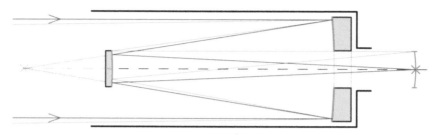

Figure 3.18. A Ritchey–Chrétien telescope. Credit: Wikipedia: Di I, ArtMechanic (CC BY-SA 3.0).

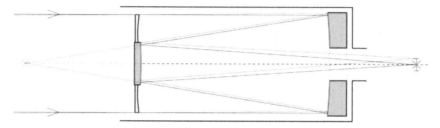

Figure 3.19. A Schmidt telescope. Credit: Wikipedia: Griffenjbs (Public Domain).

represents a compromise between aberrations and compactness, with low occultation of the secondary (Figure 3.18).

Schmidt telescope: The Schmidt telescope (1670) is a catadioptric telescope (that is, it uses reflecting and refracting parts) that consists of a spherical mirror, a correcting plate, and a curved focal plane (Figure 3.19). The fact that the primary is spherical allows even very divergent rays to have the same aberrations: it does not have a predefined axis. However, the image quality is low (but there is no coma or astigmatism). The corrector plate, of variable thickness, eliminates spherical aberration by compensating for the difference in the paths of the rays with respect to the parabolic profile. This creates a telescope with a large field of view, of as much as $5° \times 5°$.

One of the problems with most of the previous configurations is that part of the primary mirror area is obscured by the presence of secondary mirrors and receivers. In addition to obscuring objects on the telescope's line of sight, it creates diffraction problems. To overcome these problems, off-axis configurations are used in which a portion of the paraboloid is used that does not include the vertex. Off-axis telescopes are advantageous at large wavelengths where diffraction is more important (Figure 3.20).

3.4 Telescope Mounts

A telescope is supported and moved by a "mount" that must also maintain the parallelism between the axes of the different mirrors and the receiver. The mount:
- Allows for the pointing of the instrument
- Allows for the compensation of the movement due to the Earth's rotation
- Keeps any multiple of mirrors of the telescope integral and united with the receiver.

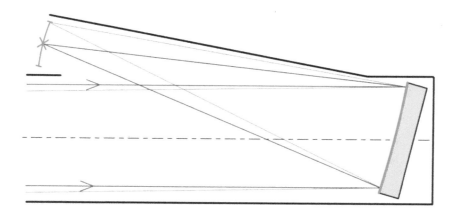

Figure 3.20. Off-axis telescope. Credit: Wikipedia: Eudjinnius (CC BY-SA 3.0).

Figure 3.21. Serrurieur Truss. Credit: Wikipedia: Krzysztof Ulaczyk (CC BY-SA 4.0).

Each mount has two degrees of freedom, two orthogonal axes around which the instrument can rotate with its motors, position sensors, and control computer. Errors in pointing must be less than 1/10 of the angular resolution. Deformations due to gravity will always be present during the movement of a telescope. The "Serrurieur Truss" is an ingenious method to have deformations that counter-balance each other, mutually maintaining alignment (Figure 3.21).

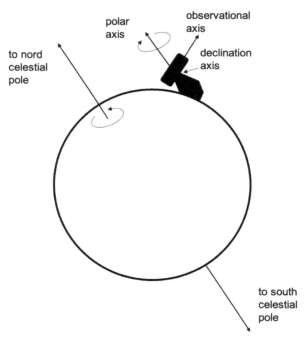

Figure 3.22. Equatorial mount. The declination axis is fixed at the declination of the celestial object, and the right ascension motor rotates around the polar axis to track the object.

There are two types of mount, an equatorial mount and an alt-azimuth mount.

Equatorial mount: In the equatorial mount, one of the axes is parallel to the Earth's rotation axis (the polar axis; Figure 3.22). This has to be oriented toward the pole (either north or south, depending on the observer's latitude) at an angle equal to the latitude of the place of observation. This axis facilitates the tracking of celestial objects (that is, the telescope stays fixed on any celestial object during its apparent motion in the sky). The right ascension movement actually allows for the rotation of the telescope around the polar axis. The second axis is that of the declination, which is fixed to the declination of the object to be observed. Tracking occurs by rotating the right ascension motor around the polar axis with constant speed. The equatorial mount must be very robust and sometimes heavy, because the polar axis is inclined with respect to the vertical, especially for medium latitudes where the polar axis is neither horizontal nor vertical. There exist two configurations. In the German equatorial mount, the declination axis is attached to the top of the right ascension motor and can accept a wide variety of telescopes. In the equatorial fork mount, a fork is attached on top of the right ascension motor. It allows for a more compact design, but has some limitations in size and weight.

Alt-azimuth mount: In the alt-azimuth mount, the axes are vertical (elevation or altitude) and horizontal (azimuth). To track an object, we always must combine the motion of the two axes with variable speed. In addition, the field observed in the sky rotates during the tracking, and therefore it often requires the presence of a

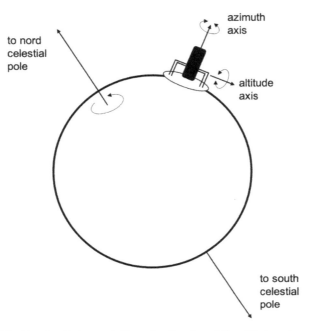

to nord
celestial
pole

azimuth
axis

altitude
axis

to south
celestial
pole

Figure 3.23. Alt-azimuthal mount. The azimuth axis and the altitude axis are shown.

derotator. When an object passes the zenith, it is impossible to keep tracking because the azimuth motor would have too high a speed, so there is always a dark region of the sky that cannot be observed (Figure 3.23).

Despite these difficulties, it is much simpler and cheaper to manufacture an alt-azimuthal mount. In fact, the axis rotation motors work on the balanced instrument. All large telescopes are alt-azimuth, with the notable exception of the 140 foot NRAO telescope in Green Bank, West Virginia (USA), whose mounting structure was built from the keel of a boat (Figure 3.24).

With an alt-azimuth mount, it is possible to place the detector (or eyepiece) at the Nasmyth focus. The Nasmyth focus is on the telescope's altitude axis and is reached by deflecting the beam with a 45° flat mirror (Figure 3.25). This allows one to insert heavy instruments that are not to be inclined at the focus of the telescope. In addition, different instruments can be mounted to the focus of the telescope that can be illuminated by simply rotating the flat mirror.

3.5 Telescopes Construction

3.5.1 Telescopes Around the World

It is clear that there are a lot of good reasons for making large telescopes. How big should they be? Optical mirrors larger than 4 m and up to 10 m are ideal for ground-based telescopes. Radio telescopes that are tens or hundreds of meters in diameter are needed to improve angular resolution. In parallel with the construction of large telescopes, observation sites where the impact of the atmosphere is lower or where

Figure 3.24. The 140 foot NRAO telescope in Green Bank, West Virginia (USA).

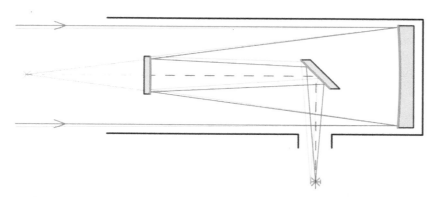

Figure 3.25. A telescope in an alt-azimuth mount with the Nasmyth focus. Credit: Wikipedia: Di Rainald62 (CC BY-SA 3.0).

Figure 3.26. Maunakea observatories in Hawaii (USA). From left to right, the United Kingdom Infrared Telescope, the Caltech Sub-Millimeter Observatory (closed in 2015), the James Clerk Maxwell Telescope, the Smithsonian Sub-Millimeter Array, the Subaru Telescope, W. M. Keck Observatory (I & II), the NASA Infrared Telescope Facility, and the Gemini North Telescope. Credit: Wikipedia: Frank Ravizza (CC BY-SA 4.0).

Figure 3.27. Left: Roque de los muchachos in La Palma (Spain). From left to right, the William Herschel Telescope, the Dutch Open Telescope, the Carlsberg Meridian Telescope, the Swedish Solar Telescope, the Isaac Newton Telescope, and the Jacobus Kapteyn Telescope. Credit: Wikipedia: De H. Zell, Trabajo propio (CC BY-SA 3.0). Right: Teide Observatory in Tenerife (Spain). Credit: Wikipedia: De Besnier.m—Trabajo propio (CC BY 2.5).

the precipitable water vapor (PWV) is negligible began to be exploited. Among them we mention the following:

- Maunakea observatories on the Big Island in Hawaii (USA; Figure 3.26)[2]
- Roque de los muchachos and Teide observatory in the Canary Islands (Spain; Figure 3.27)[3]
- South Pole station and Dome C station in Antarctica[4,5]
- The Atacama desert (Chile; Figure 3.28)
- Paranal[6] and La Silla[7] on the Andes mountains (Chile)
- Green Bank Observatory, West Virginia (USA)

---

[2] http://www.ifa.hawaii.edu/mko/

[3] http://www.iac.es/

[4] https://en.wikipedia.org/wiki/Amundsen%E2%80%93Scott_South_Pole_Station

[5] https://en.wikipedia.org/wiki/Concordia_Station

[6] https://www.eso.org/public/teles-instr/paranal-observatory/

[7] https://www.eso.org/public/chile/teles-instr/lasilla/

Figure 3.28. Telescopes in the Atacama desert (Chile). Left: an artist's view of the Atacama Large Millimeter/submillimeter Array (ALMA). Credit: Wikipedia: ALMA (ESO/NAOJ/NRAO)/L. Calçada (ESO; CC BY 4.0). Right: the Atacama Cosmology Telescope at 5200 m above sea level. Credit: Wikipedia: Uploaded by Ahincks (CC BY 3.0).

Figure 3.29. Left: the European Southern Observatory (ESO) Paranal observatory (Chile). The Very Large Telescope (4 units) and the VLT Survey Telescope can be seen. Credit: Wikipedia: Rivi~commonswiki assumed (CC BY-SA 3.0). Right: ESO La Silla observatory (Chile), seen from the NTT telescope. Credit: Wikipedia: De Hernan Fernandez Retamal - Trabajo propio (CC BY-SA 3.0).

- Noto, Medicina, and Sardinia Radio Telescopes (Italy)
- The Very Large Array in New Mexico (USA; Figure 3.29)[8]
- The Australian Telescope Compact Array (ATCA) and the Parkes radio telescope (Australia)[9]
- The Institut de Radioastronomie Millimétrique (IRAM, Spain).[10]

3.5.2 Quality of the Mirrors

Atmospheric effects do not allow us to take advantage of optical telescopes larger than a few meters, which, in any case, are very difficult to manufacture. The primary mirror of a telescope is affected by small-scale (roughness) and large-scale deformations. A statistical analysis of the roughness shows that the rms of the mirror surface

---

[8] https://it.wikipedia.org/wiki/Very_Large_Array
[9] https://en.wikipedia.org/wiki/CSIRO
[10] https://www.iram-institute.org/

fluctuations disperses the radiation onto a non-negligible area that results in a loss of efficiency f:

$$f \approx 1 - e^{\frac{4\pi\sigma_{rms}}{\lambda}} \approx \frac{4\pi\sigma_{rms}}{\lambda} \qquad (3.11)$$

which means that the rms of the surface fluctuations must be $\ll\lambda$ for the efficiency to approach 1.

3.5.3 Deformation of the Mirror

Large-scale deformations are largely determined by gravity: they can be reduced using honeycomb manufacturing techniques such as for the Hubble Space Telescope and/or with mirrors obtained by fusing the glass in rotating furnaces that are then turned off and kept rotating (Figure 3.30). Also, a recent manufacturing technique makes a liquid mirror telescope with mercury or with gallium. Examples of these telescope are the Canadian Large Zenith Telescope[11] or the James Webb Space Telescope.[12] The Palomar Hale telescope in California (1948, 5 m, 15 t)[13] has long been the largest telescope in the world. The mirrors deform under their weight.

Figure 3.30. Image of the mirror of the Hubble Space Telescope, uncoated. This shows the honeycomb structure of the primary mirror that is used to be lightweight and have structural resistance. Credit: Wikipedia: Levg (Public Domain).

---

[11] https://www.astro.ubc.ca/lmt/lzt/

[12] https://www.jwst.nasa.gov/

[13] https://www.jpl.nasa.gov/images/hale-telescope-palomar-observatory

A solution used since the 1980s is active optics: actuators maintain the shape of the mirrors (which are necessarily thin) while the telescope moves, if there are thermal deformations and if the secondary is decentralized during a track. The actuators are controlled by a calculator that analyzes the image of a star and calculates the actuator positions. This technique is necessary for telescope of ~8 m diameters, and is used by, among others:

- Telescopio Nazionale Galileo (La Palma, Spain)[14]
- New Technology Telescope (La Silla, Chile)[15]
- Keck Telescope (Maunakea, Hawaii, USA)[16]
- Nordic Optical Telescope (La Palma, Spain)[17]
- Gran Telescopio Canarias (La Palma, Spain; Figure 3.31).[18]

Active optics are also essential for mosaic telescopes that allow one to relax the required tolerances. These telescopes are easier to manufacture and are more lightweight. One of the first of these telescopes was the Multi Mirror Telescope (6×1.8 m mirrors combinable in different configurations)[19].

Figure 3.31. Actuators for the active optics of the Gran Telescopio Canarias (La Palma, Spain). Credit: Wikipedia: H. Raab (User:Vesta)—(CC BY-SA 3.0).

[14] http://www.tng.iac.es/

[15] https://www.eso.org/sci/facilities/lasilla/telescopes/ntt.html

[16] https://www.keckobservatory.org/

[17] http://www.not.iac.es/

[18] http://www.gtc.iac.es/

[19] http://www.mmto.org/

Figure 3.32. Sardinia Radio Telescope (Italy), an active surface telescope. Credit: Wikipedia: Di caprowsky (CC BY 3.0).

A similar technique is the one used for large radio telescopes. In this case, it is called an "active surface". Examples of these types of telescopes are:

- Sardinia Radio Telescope (Italy; Figure 3.32)[20]
- Green Bank Telescope (West Virginia, USA)[21]
- Large Millimeter Telescope (Mexico)[22]
- Noto Radio Observatory (Italy).[23]

3.5.4 Adaptive Optics

In the visible band, one of the effects that most hampers the quality of an observation is the effect of the atmosphere (Figure 3.33). When a plane wave passes through the atmosphere, it gets distorted by the refractive index of the atmosphere that is variable in time. This affects the quality of the observations and can be quantified by the *seeing*, which is a measure of the dispersion of a point-like source in the sky and can be measured as an angle. The order of magnitude of the seeing at good observational sites ranges from 0.5″ to 2″. Seeing can be improved using adaptive optics.

Atmospheric fluctuations are monitored in real time through the use of a sodium laser and is based on the radiation reflected back from the upper layers (90 km) of the atmosphere. This monitoring allows for the calculation of the deformation of a secondary or tertiary mirror, which compensates for the deformations and which minimizes the size of a point source in the sky. This is a typical example of a

---

[20] http://www.srt.inaf.it/

[21] https://greenbankobservatory.org/science/telescopes/gbt/

[22] http://lmtgtm.org/

[23] https://www.noto.ira.inaf.it/

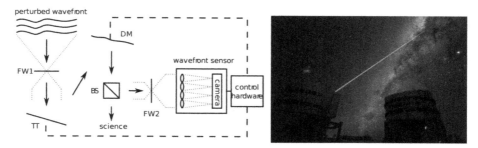

Figure 3.33. Left: an adaptive optics scheme. The perturbed wave front first hits a tip–tilt (TT) mirror and then a deformable mirror (DM), which corrects the wave front. Part of the light is tapped off by a beamsplitter (BS) and sent to the wave front sensor and the control hardware, which sends updated signals to the DM and TT mirrors. Credit: Wikipedia: Yuri Beletsky—NASA Astronomy Picture of the Day (CC BY 4.0). Right: a laser is used to measure the deformation to be applied. Credit: Wikipedia: 2pem (CC BY-SA 3.0).

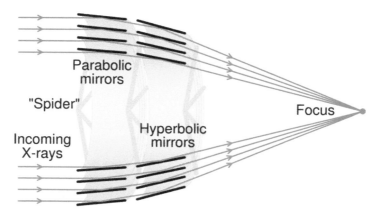

Figure 3.34. A grazing incidence telescope. Credit: Wikipedia: Cmglee (CC BY-SA 4.0).

feedback loop that can be processed with a proportional, integral, and differential (PID) algorithm (see Section 7.4).

Clearly, at large wavelengths, there are different problems. The roughness of a mirror surface is a less of a fundamental problem in the millimeter band and at radio frequencies. However, angular resolution worsens due to diffraction. Infrared detectors have been developed only recently, and IR telescopes have the problem of reducing their own emission. One has to be careful of the secondary lobes and under-illuminate the mirrors. Some techniques mitigate the problem of telescope and atmospheric emissions by means of spatial modulation, by alternately observing the source and a surrounding area that is free of sources. For example, a Cassegrain telescope can have a secondary oscillating mirror that allows you to "illuminate" an area of sky adjacent to that of interest. Obviously, the mirror must be light (e.g., made of carbon fiber).

3.6 X-ray Telescopes

Unlike an optical telescope that uses materials with a refractive index different from $n = 1$ and for which the incident light rays are almost normal, in the X-ray band it is not possible to use the same materials, due to the penetrating power of the X-rays. The X-rays are transmitted or absorbed by the mirrors. An X-ray telescope is instead a grazing incidence (or Wolter) telescope, and sometimes has a nested configuration (Figure 3.34). This type of design was invented by Hans Wolter in 1952 in three configurations. Disadvantages include the fact that the field of view is very small.

Experimental Astrophysics

Elia Stefano Battistelli

Chapter 4

Signal Processing

The basics of the theory of signals are presented in this chapter. The concepts of Fourier transforms, power spectra, autocorrelation functions, and the Wiener–Khinchin theorem are introduced. Transmission lines, the difference between analog and digital instrumentation, the basic working principles of an analog-to-digital converter (ADC) and of a digital-to-analog converter (DAC), sampling and quantization, aliasing, and the Nyquist–Shannon theorem are demonstrated here.

4.1 Fourier Transform

Electronics deals with the use and processing of electrical signals as carriers of information (astrophysical and beyond). In many applications, an astrophysical signal is transduced into an electrical signal, which in turn is transformed into a visual signal on a monitor or on a hard disk. In signal theory we focus on the correct use and interpretation of electrical signals that travel through our instruments and on their preservation. In electronics, a signal describes a variation of electric current within a conductor: clearly, a signal cannot be represented by a direct current that would not carry information after the first measurement. Any information included in a S signal is also accompanied by noise N. A quantity of fundamental importance in signal theory is the signal-to-noise ratio S/N, the measure of how much a signal is stronger than the noise that accompanies it.

A periodic function over time can be expressed as the sum of periodic functions (sines and cosines) at different frequencies. Given a periodic function $f(t)$, it can be expressed as a Fourier series as in the following:

$$f(t) = \frac{1}{2}a_0 + \sum_n [a_n\cos(n\omega t) + b_n\sin(n\omega t)] \tag{4.1}$$

doi:10.1088/2514-3433/ac0ce4ch4

where:

$$a_n = \frac{1}{T} \int_0^T f(t)\cos(n\omega t)dt$$

$$b_n = \frac{1}{T} \int_0^T f(t)\sin(n\omega t)dt$$

with the period of the function defined as:

$$T = \frac{2\pi}{\omega} = \frac{1}{f}.$$

Using the complex formalism:

$$f(t) = \sum_{n=0}^{\infty} c_n e^{in\omega t} \qquad (4.2)$$

where:

$$c_n = \frac{1}{T} \int_{-T/2}^{T/2} f(t)e^{-in\omega t}dt.$$

If the function is not periodic, it can still be considered as a periodic function with infinite period. Consequently, the $n\omega$ frequency will become infinitesimal, and we can pass to the continuum instead of performing a discrete summation. Considering the exponential form of sines and cosines, we can write:

$$f(t) = \frac{1}{2\pi} \int_{-\infty}^{\infty} g(\omega)e^{i\omega t}d\omega \qquad (4.3)$$

and

$$g(\omega) = \int_{-\infty}^{\infty} f(t)e^{-i\omega t}dt \qquad (4.4)$$

where $f(t)$ is the inverse Fourier transform and $g(\omega)$ is the Fourier transform of the function $f(t)$, i.e., the spectrum of $f(t)$.

4.2 Transmission Lines

A transmission line is a couple of conductors in a circuit that transmits a signal, a voltage or a current, from one point to another (Figure 4.1). In small circuits, the fact that the propagation speed of electrical signals is not infinite can be neglected. For small circuits, and for signals with limited frequencies (<MHz), this approach is correct. Considering that a signal propagates at speed $\sim c$ in a 30 cm circuit, the delays are on the order of nanoseconds.

A ladder line is a couple of conductors that transmit a low-frequency signal. If the signal to be transmitted has high frequencies, such as GHz, and has to travel a long distance, typical pulses ($\sim\mu$s) will be greater than the period of the signals. In order to transmit signals of this type in a controlled way, a transmission line is used that

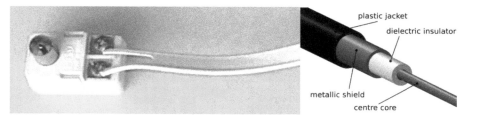

Figure 4.1. Transmission lines. Left: a ladder line for low-frequency signals. Right: a coaxial cable. Credit: left: Wikipedia: Tkgd2007 (CC BY 3.0); right: Wikipedia: Andrew Alder (CC BY-SA 3.0).

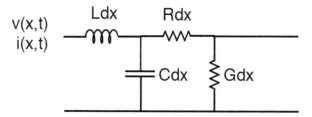

Figure 4.2. Transmission line schematic.

consists of two conductors that allow a signal to be transmitted while keeping under control the capacitive and inductive effects distributed along their length: a coaxial cable. A coaxial cable is characterized by a bandwidth (a maximum frequency that can be transmitted with no degradation). An example of a transmission line is a coaxial cable composed of a copper conductor and a cylindrical metallic shield separated by a dielectric (typically polyethylene). We will have to consider the distributed inductance L of the cable, as well as the distributed capacity C between the two cables. Assuming that these quantities are per unit of length dx, we can draw the coaxial cable as in Figure 4.2.

Assuming that the line is "non-dissipative", we can neglect the resistance R and the conductance G. So, we will have a voltage and a current as in the following:

$$dV = -Ldx\frac{\partial i}{\partial t} \rightarrow \frac{\partial V}{\partial x} = -L\frac{\partial i}{\partial t} \tag{4.5}$$

$$dI = -Cdx\frac{\partial V}{\partial t} \rightarrow \frac{\partial I}{\partial x} = -C\frac{\partial V}{\partial t} \tag{4.6}$$

and so:

$$\frac{\partial^2 V}{\partial x^2} = -L\frac{\partial}{\partial t}\frac{\partial i}{\partial x} = LC\frac{\partial^2 V}{\partial t^2} \tag{4.7}$$

$$\frac{\partial^2 I}{\partial x^2} = -C\frac{\partial}{\partial t}\frac{\partial V}{\partial x} = LC\frac{\partial^2 I}{\partial t^2} \tag{4.8}$$

which is the wave equation of a wave that propagates with speed:

$$v = \frac{1}{\sqrt{LC}}. \tag{4.9}$$

That said, $V = Zi$, with characteristic impedance $Z = \sqrt{\frac{L}{C}}$.

The capacity of a cylindrical capacitor with outer and inner diameters D and d, from the Biot–Savart law, are:

$$C = \frac{2\pi\varepsilon_0\varepsilon_r}{\log\left(\dfrac{D}{d}\right)} \tag{4.10}$$

and its inductance is

$$L = \frac{\mu_0\mu_r}{2\pi}\log\left(\frac{D}{d}\right). \tag{4.11}$$

Thus, the speed of the wave is the following:

$$v = \frac{1}{\sqrt{LC}} = \frac{1}{\sqrt{\mu_0\mu_r\varepsilon_r\varepsilon_0}}. \tag{4.12}$$

Usually, the magnetic permeability of the medium is that of the vacuum and the relative dielectric constant is ~2. Typical values for capacity and inductance are $C \sim$ 100 pF m^{-1} and $L \sim$ 250 nH m^{-1}, so we have:

$$v = \frac{c}{\sqrt{\varepsilon_r}} \sim 0.7c \tag{4.13}$$

$$Z = \sqrt{\frac{L}{C}} \sim 50\,\Omega \tag{4.14}$$

which is the typical impedance of commonly used transmission lines. Checking the impedance of a transmission line is important to "match" it with the output and input impedance of the instruments used in order to maximize the power transferred.

Power supply lines in Europe are at 50 Hz and 220 V. These produce a variable electric field that couples capacitively with any conductor, which in turn produces a current in it. This depends on the grounding level at all frequencies. The same currents create magnetic fields that couple inductively with any circuit. Magnetic fields are weaker (if one does not use AC motors, such as vacuum pumps, etc.), but they are also more difficult to shield. When (non-coaxial) stand-alone wires are used, the cables are usually twisted to be subject to opposite electrical interferences between the wires crossed by opposite currents and to reduce the effective area presented to the magnetic interference fields. In addition, the shielding of a pair of cables allows for the cable to be isolated through a Faraday cage. In a coaxial cable, the external shield is not only the ground reference, but also acts as a Faraday cage (a third shield is introduced in triaxial cables).

4.3 Digital versus Analog Signals: ADC and DAC

A physical measurement can be made by direct comparison with a sample, or through the use of a measuring instrument. The latter must be calibrated. Among the features of an instrument are:

- The dynamic range
- The sensibility
- Time constant
- Accuracy.

At the base of a measuring instrument there is always a sensor: a device that transduces one physical (or electrical) quantity into another. Measuring instruments can be either analog or digital. An analog instrument is based on the use of (or made of) springs, expanding liquids, amplifiers, oscillators, modulators, etc., which allows a signal to be "read" by comparing it with a standard unit. A digital instrument, however, has a sensor and an analog-to-digital converter (ADC) with n bits, which converts a voltage into a binary number. Also, a digital instrument has logic gates, microprocessors that process the read data. In addition, a digital instrument usually has a display.

When comparing an analog instrument with a digital instrument, we highlight the following differences (some are approximate): an analog instrument is usually more complicated to use, more expensive, less versatile, and it cannot do an autocalibration or save data. It has a possible reading error, and noise usually degrades the signal as well as its transmission, compared with a digital instrument. That said, a digital instrument has two important drawbacks: the signal is a discrete one (see Figure 4.3), and the bandwidth is limited. If these two effects are well

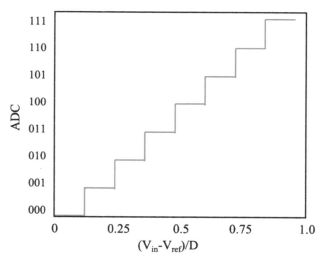

Figure 4.3. A 3 bits ADC conversion. The base reference (V_{ref}) was removed from the input signal (V_{in}), and the result was divided by the dynamics D of the ADC. Clearly, the conversion implies a rounding approximation.

understood, there is no problem in using a digital instrument, but, they can give rise to quantization and a sampling problem that consequently aliases the signal.

More and more of everyday life involves the use of digital signals. An analog signal is, for example, an electrical signal that varies with both spatial and temporal continuity within its valid range. A digital signal is, instead, a signal that varies discontinuously, in a quantized way, within its N possible values. In our instruments, therefore, a conversion from an analog signal to a digital signal and vice versa has to be accounted for. The decimal number system we use is based on the powers of 10, which is an "anthropic" and arbitrary choice. In electronics, the operating basis of a digital system is the switches; therefore, a numerical system with base 2 is used. A digital signal is therefore a binary number with n bits: a sequence of n numbers, either 0 or 1. In this approximation, an n bits instrument can read N different numbers, where:

$$N = 2^n \tag{4.15}$$

$$0 \leqslant N \leqslant 2^n - 1. \tag{4.16}$$

4.3.1 DAC

A digital-to-analog converter (DAC) is an electronic component capable of converting a digital signal into an analog (current or voltage) signal. The conversion table is called a *Look Up Table* (*LUT*). Suppose we have a digital (binary) number of digits (V_i), or N bits, and we want to convert it into an analog signal. The output signal must be proportional to it. We can use an "analog adder", which is an application of an amplifier, the summing amplifier, that carries out one conversion with a "weighted" DAC:

Referring to Figure 4.4, we have that:

- $V_i = 0$ means that the switch is open (toward ground).
- $V_i = 1$ means that the switch is closed (towards a reference voltage V_{ref}).

In this n bits DAC, the resistors have values of $R_n = R$, $R_{n-1} = R/2$, $R_{n-2} = R/4$, ..., and each V signal can be connected to a reference voltage V_{ref} depending on if V_i

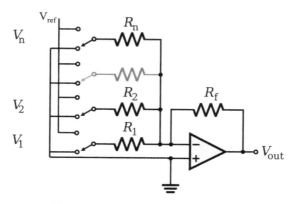

Figure 4.4. A digital-to-analog converter.

is 1 or 0. The nth is the most significant bit (MSB), and the 0th is the least significant bit (LSB). The digital starting signal is thus:

$$V_0^{\text{dif}} = \sum_{i=0}^{n-1} V_i \cdot 2^i = V_0 \cdot 2^0 + V_1 \cdot 2^1 + V_2 \cdot 2^2 + \ldots + V_n \cdot 2^n. \qquad (4.17)$$

The analog signal will therefore be:

$$
\begin{aligned}
V_0^{\text{an}} = I \cdot R &= (I_0 + I_1 + I_2 + \ldots + I_n) \cdot R \\
&= R \cdot V_{\text{ref}}\left(\frac{V_0}{8R} + \frac{V_1}{4R} + \frac{V_2}{2R} + \ldots + \frac{V_n}{R}\right) \qquad (4.18) \\
&= \frac{R}{4R} \cdot V_{\text{ref}}(V_0 \cdot 2^0 + V_1 \cdot 2^1 + V_2 \cdot 2^2 + \ldots + V_n \cdot 2^n)
\end{aligned}
$$

which is indeed a voltage whose value is proportional to the input digital number.

The linearity of this conversion is linked to the precision of the resistors R; the dynamics is instead linked to the value of the resistors. The parameters of a DAC are related to the ideal transfer characteristic and are:

- Resolution: This is the minimum possible variation of the output signal; it is equal to the output voltage of the LSB.
- Offset: This arises in cases where the output is not null, even if the input is null.
- Gain error: This is the difference between a theoretical maximum and the real value.
- Linearity error: This is the difference between the real and ideal transfer characteristic.
- Settling time: This is the time taken to complete the conversion.

4.3.2 ADC

An analog-to-digital converter (ADC) is an electronic component that does the opposite operation to that a DAC: it converts an analog signal such as a current or voltage into a binary number of n bits. The basic operation within an ADC is the comparison of the analog signal with a linearly increasing digital signal. The conversion ends when the difference between the two signals is less than a predetermined threshold (usually the resolution of the ADC or the minimum detectable signal or LSB). Direct comparison has the disadvantage of generating the digital signal with times that are dependent on the input signal. The most common method is that of the successive approximation, which is independent of the input signal. In the successive approximation method, the analog signal is constantly compared with a gradually smaller digital signal. An ADC is composed of the following components (see Figure 4.5):

- A "comparator" compares the signal V_{IN}, after a sample-and-hold circuit (S/H), with a signal generated by the DAC V_{D}, and gives 0 if $V_{\text{D}} > V_{\text{IN}}$ or 1 (actually a reference voltage V_{C}) if $V_{\text{D}} < V_{\text{IN}}$.
- A successive approximation register (SAR), in correspondence with each clock pulse, transfers the logical level of V_{C} on one output line at a time, starting from

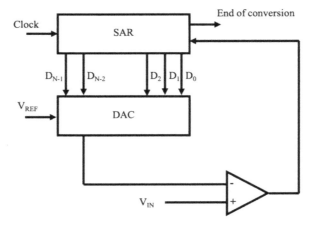

Figure 4.5. A successive approximation analog-to-digital converter.

the MSB, D_{N-1}, and continuing with the immediately lower-weight bit to the LSB, D_0. This approximate signal is then transferred to a DAC.

- A DAC transfers the signal from the SAR to the comparator and also iterates from the MSB to the LSB.
- A viewer o exit register has the task of storing the digital data D_i of the output of the SAR and makes it available to the user.

An ADC has the following characteristics:
- Resolution: a number of possible discrete values, measured in bits (e.g., 8 bits corresponds to 256 values, from 0 to 255)
- Conversion range: the voltage range that can be converted (e.g., 0 V–5 V)
- Conversion time: the time taken to answer, which depends on the clock and on the number of bits
- Quantization error: the voltage V relative to the LSB, a quantity linked to the resolution and the conversion interval
- Linearity error: the difference between the characteristics of ideal and real transfer.

4.4 Quantization

A digital signal has the ability to reject noise; an analog signal is a continuous function on which noise is superimposed. When a signal is digitally converted (when correctly interpreted), it is interpreted as the closest number, even when noise overlaps it (Figure 4.6). A digital signal can be processed more easily and with higher computing power; on an analog signal, only a few operations can be done with adder or integrator circuits, but the operations are difficult and limited. A digital signal can be faithfully recorded; techniques such as magnetic tapes and paper recorders are used for an analog signal and can have some limitations. Given all these advantages, one may think that digital instruments are undoubtedly better. However, there are two main disadvantage which should be addressed:

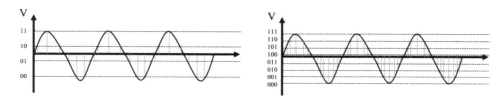

Figure 4.6. Digital conversion with 2 (left) or 3 (right) bits.

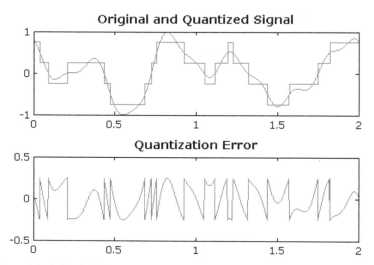

Figure 4.7. Top: original signal (blue) and digitized signal (red). Bottom: the difference between the original and digitized signals, the quantization error. Credit: Wikipedia: Atropos235 (Public Domain).

- A digital signal is always quantized: the digital signal does not vary continuously within the validity range, but varies by step. The number of possible steps N is linked to the number of bits of the signal n.
- A digital signal is sampled: in order to convert a signal from analog to digital, a signal is not "picked up" continuously, but at consequent instants of time. During a conversion, not all of the infinite possible values of the analog continuous function can be converted to digital. Quantization is an irreversible, and clearly non-linear, "rounding" process.

This could lead to a loss of information and generate additional noise that will at most be equal to half the LSB. The best way to address this problem is to increase the number of bits in an ADC (Figure 4.7).

When the capacity is the same as that of an ADC (of n bits), whether quantization noise is acceptable or not must be decided *a priori*, based on the signal to be detected and its intrinsic noise. In terms of the signal-to-noise ratio, S/N, if a signal V_s "takes" all of the dynamics of the ADC, $V_s = V_{max}$, then we have:

$$S \approx V_{max} = 2^n \cdot \frac{V_{max}}{2^n} \qquad (4.19)$$

$$N = \frac{V_{\max}}{2^n} \tag{4.20}$$

$$\frac{S}{N} \approx 2^n. \tag{4.21}$$

If instead the signal V_s "occupies" only part (m bits) of the dynamics of the ADC, then we have:

$$S \approx V_{\max} = 2^n \cdot \frac{V_{\max}}{2^{n+m}} \tag{4.22}$$

$$N = \frac{V_{\max}}{2^n} \tag{4.23}$$

$$\frac{S}{N} \approx 2^{n-m}. \tag{4.24}$$

If $m \sim n$, this is not a good thing. In order to decide the number of bits one needs, we must compare the quantization noise N_{qu} with the intrinsic noise N_{sig} of the signal, and choose the number of bits that corresponds to $N_{\mathrm{qu}} < N_{\mathrm{sig}}$.

4.5 Sampling: Nyquist–Shannon theorem

Given a variable analog signal, in order to convert it into a digital signal we have to "measure" it at regular times. This sampling can result in the loss of some of the information carried by the analog signal (Figure 4.8). Qualitatively we can say that in order not to lose information, we have to sample quickly. The question is: how fast?

In analog electronics, two different signals always appear different, but in digital electronics, two different signals may look the same due to bad sampling: this is the problem of aliasing. Here, we will deal with temporal aliasing, but there is also spatial aliasing. Suppose we sample a sinusoidal signal at an "inadequate" frequency

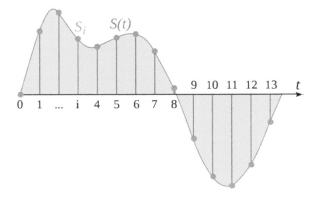

Figure 4.8. An analog signal (green) sampled at subsequent times (vertical lines). Credit: Wikipedia: Д.Ильин: vectorization (CC0).

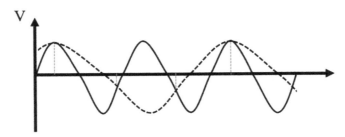

Figure 4.9. A sinusoidal signal and its (low-frequency) sampling. The new sampled signal has a totally different frequency with respect to the original signal.

lower than that of the signal itself: the frequency of the measured signal will be grossly wrong (see Figure 4.9).

So, to sample a periodic signal, we cannot use too low of a sample rate (compared to the signal frequency). The sampling theorem, or Nyquist–Shannon theorem, defines this minimum frequency. Given a superposition of signals at different frequencies (of which $f_{max} = B$ is the maximum frequency, or bandwidth), the minimum sampling frequency $f_s = 1/\Delta t$ necessary to avoid aliasing (and a loss of information) is twice the frequency f_{max}. This needs to be $f_s = 2f_N > 2f_{max}$, where f_N is defined as the Nyquist frequency.

Let P be the sampling function that samples a signal every Δt. Then:

$$p(t) = \begin{cases} P_n \text{ for } t = n\Delta t \\ 0 \text{ for } t \neq n\Delta t \end{cases}. \tag{4.25}$$

We have that:

$$p(t) = \sum_{n=-\infty}^{\infty} \delta(t - n\Delta t) = \sum_{n=-\infty}^{\infty} \delta\left(t - n\frac{n}{f_s}\right) = \sum_{n=-\infty}^{\infty} c_n e^{i2\pi n f_s t} \tag{4.26}$$

where:

$$c_n = \frac{1}{\Delta t} \int_{-\frac{\Delta t}{2}}^{\frac{\Delta t}{2}} \delta(t - n\Delta t) \cdot e^{-i2\pi n f_s t} dt = \frac{1}{\Delta t} \int_{-\frac{\Delta t}{2}}^{\frac{\Delta t}{2}} \delta(t - n\Delta t) \cdot e^{-i2\pi \frac{n}{\Delta t} t} dt$$

$$= \frac{1}{\Delta t} e^{-i2\pi \frac{n}{\Delta t} 0} = \frac{1}{\Delta t} = P_n. \tag{4.27}$$

We thus have

$$p(t) = \sum_{-\infty}^{\infty} P_n e^{i2\pi n f_s t}. \tag{4.28}$$

The sampled signal $x_A(t)$ is then:

$$x_A(t) = x(t) \cdot p(t) = \sum_{n=-\infty}^{\infty} P_n x(t) e^{i2\pi n f_s t}. \tag{4.29}$$

The Fourier transform of the sampled signal is therefore:

$$X_A(f) = \int_{-\infty}^{\infty} x_A(t)e^{-i2\pi ft}dt = \int_{-\infty}^{\infty} \sum_{-\infty}^{\infty} P_n x(t)e^{i2\pi nf_s t} \cdot e^{-i2\pi ft}dt$$

$$= \sum_{-\infty}^{\infty} P_n \int_{-\infty}^{\infty} x(t)e^{-i2\pi (f-nf_s)t}dt. \tag{4.30}$$

However, the Fourier Transform of the original signal is:

$$X(f) = \int_{-\infty}^{\infty} x(t)e^{-i2\pi ft}dt \tag{4.31}$$

thus:

$$X_A(f) = \sum_{-\infty}^{\infty} P_n X(f - nf_s). \tag{4.32}$$

So, the Fourier transform of a sampled signal is the Fourier transform of the original signal repeated at multiples of the sampling frequency (modulated by the P_n factors).

This is graphically interpreted in Figure 4.10:

So, given an astrophysical signal, it will contain information that, for example, depends on the time constant of the detectors and on the scanning strategy. This

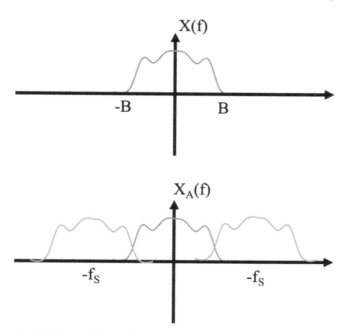

Figure 4.10. Top: the Fourier transform of the original signal $X(f)$ drawn in blue. Bottom: the Fourier transform of the sampled signal $X_A(f)$ (blue–green). When the function is sampled, at rate f_s, a repetition of the same transform is repeated at multiples of f_s. If $f_s < 2B$, then there is a range of frequencies where the actual frequency is mirrored with respect to $f_s/2$.

information will be characterized by a bandwidth of the signal f_{max}. In order to sample this signal, we cannot use too low a sampling rate (compared to the frequency of the f_{max} signal). The sampling theorem defines this minimum frequency (the Nyquist frequency): *the minimum sampling frequency necessary to avoid aliasing (and loss of information) is twice the frequency f_{max} present in the signal*. In addition to this warning, the sampling theorem (or Nyquist–Shannon theorem) also provides guidance, in that it sets an adequate sampling frequency and it is not necessary to have a sampling frequency higher than that. If this is not done, there is a range of frequencies that get mirrored with respect to the $f_s/2$.

Considering the noise of the signal while it is transported, if an astrophysical signal is present up to a frequency f_{max}, the noise can be added, before conversion, to all frequencies even higher than f_{max} so, before the conversion, it is necessary to add a filter, an anti-alias low-pass filter, which removes the noise above f_{max}.

Experimental Astrophysics

Elia Stefano Battistelli

Chapter 5

Semiconductor Physics

This chapter is dedicated to semiconductors, doped semiconductors, junctions, diodes, photomultipliers, and the use of all of these in astrophysics. We will discuss junction-gate field-effect transistors, transistors, operational amplifiers, and differential amplifiers, as well as their use.

5.1 Semiconductors and Junctions

The difference between a metal and an insulator is clearly connected to their electrical resistance. A metal is a material with a very low ionization energy, so many electrons separate from their respective atoms and are free to move. By applying an electric field to a metal, the electrons accelerate, but due to impacts with the lattice, they maintain a constant speed: this is the source of electrical resistance in metals. An insulator, however, has, virtually, an infinite electrical resistance. In semiconductors the ionization energy is much higher than in metals. At temperatures close to absolute zero, there are no free electrons, but at room temperature some bonds break due to thermal ionization. In this case, there are free electrons, creating electrons n and holes p (which attract and repel neighbors' electrons, and therefore "move" too).

Among semiconductors, the most widely used are silicon and germanium. For silicon we have a gap energy, which is necessary to ionize the material, of $E_G = 1.12$ eV. At ambient temperature, the concentration $n = p = n_i \sim 1.5 \times 10^{10}$ carriers per cm^3 (to be compared with an atomic density of 10^{22} per cm^3), which is a very low value. In order to increase the number of carriers, one has to input into the semiconductor an element close to it in the periodic table that increases them: a doping element. A doped semiconductor does not have $n = p$. If a pentavalent impurity or a trivalent element is added to the silicon (which is tetravalent), an electron or a hole will be free. This will increase the number of carriers. By doping a semiconductor, it is possible to have semiconductors with charge carriers almost exclusively positive (p) or negative (n; Figure 5.1).

doi:10.1088/2514-3433/ac0ce4ch5

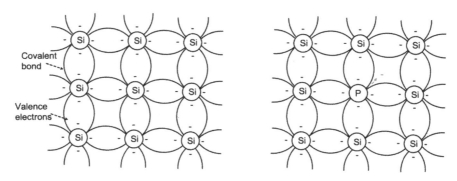

Figure 5.1. Left: tetravalent (silicon) semiconductor. Right: silicon doped with phosphorus (pentavalent).

Imagine creating a p–n junction by welding a doped crystal p and a doped crystal n (or by doping a single crystal differentially in two parts of the crystals itself). Attracted by the different concentrations of charge carriers, some holes migrate from p to n and recombine with electrons, and vice versa in the opposite direction, giving rise to a diffusion current I_D. This is a diode. A depletion region is created in which electrons and holes neutralize, leaving atoms that are no longer neutral (positive in n and negative in p). The double layer that is created generates an electric field p–n (and therefore a potential difference V) that opposes the diffusion, creating a state of equilibrium. The balance is dynamic with an I_S current generated by the potential barrier, which compensates for I_D (Figure 5.2).

If we connect a p–n junction to a voltage generator V_R, which would tend to make current flow in the direction n–p, we have a reverse bias. The external potential is added to the existing barrier; therefore, it decreases I_D until it is canceled out, while I_S remains constant. The effect is that of increasing the voltage barrier. There will be a small net current independent of V_R. However, if V_R exceeds a breakdown threshold, the junction changes behavior abruptly. The applied electric field breaks the covalent bonds, electron–hole pairs are generated, and a strong current is generated. In this situation, the charge carriers can in turn break the covalent bonds in the crystal, triggering an avalanche process, which, however, may not be destructive (Figure 5.3).

If we connect a p–n junction to a voltage generator V_F that would tend to make current flow in the direction p–n, we have a forward bias. The external potential reduces the existing barrier; therefore, I_D increases while I_S remains constant. The barrier decreases and when the applied electric field cancels the potential barrier, there is a strong passage of charges and the junction behaves like a very small resistance (with an exponential characteristic; Figure 5.4).

5.2 Diodes

As already mentioned, a component made by means of a p–n junction, for example of silicon, is a diode. The current–voltage, $I–V$, characteristic is clearly non-linear and is shown in the following figure. In the forward direction (forward bias) the current can be considered equal to zero up to a threshold value V_d. Above V_d, the current increases suddenly and its value is dictated by the circuit in which the diode

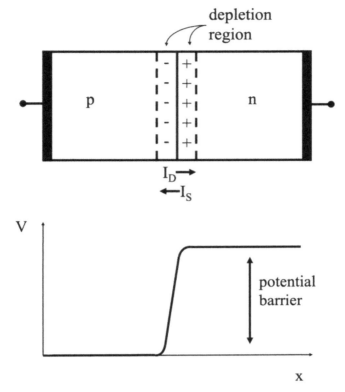

Figure 5.2. Top: diode junction. Bottom: potential barrier of the same junction.

itself is placed: at high direct voltages, the diode is similar to a short circuit. Every diode has a maximum current beyond which the physical breakdown of the component occurs. In reverse bias, the current is minimal up to the breakdown voltage (which, however, may not be dangerous). The *I–V* characteristic (or Shockley diode equation) can be expressed as the following:

$$I = I_S\left(e^{\frac{qV_D}{nk_BT}} - 1\right)$$ (5.1)

where k_BT/q is the thermal voltage, about 26 mV at ambient temperature, and n is the diode ideality factor, which is from 1 to 2 for silicon. The *I–V* characteristic has a strong dependence on the temperature (Figure 5.5). This is why semiconductors are often used as radiation detectors like bolometers (see the next chapter).

A diode is usually schematized and modeled as a bipolar component that allows current to pass only in one direction (Figure 5.6). Resistance is zero in forward bias and infinite in reverse bias.

Another option is shown in Figure 5.7. Considering the following circuit, the response to a DC voltage will be a straight line if forward biased or zero if reversed biased.

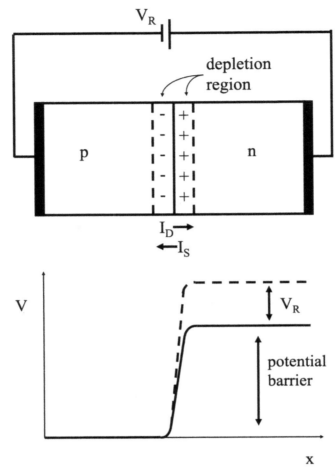

Figure 5.3. Reverse biased diode junction. Top: diode junction. Bottom: potential barrier.

Another application of diodes is to use them as clippers (Figure 5.8). In the output side circuits, we have: $V_0 > V_R - V_d$.

In the case of a sinusoidal voltage, diodes can be used as rectifiers (Figure 5.9).

In reality, the signal obtained has a positive average, but it is not exactly constant. If we add a low-pass filter, we get a rectifier (Figure 5.10).

In general, diodes are manufactured with different materials and modes to be used for different observations and with different polarizations. Some are forward biased, others are reverse biased (and even used in breakdown conditions). The most used diodes in astrophysics are listed in the following and are detailed in the subsequent paragraph:

- Zener diode
- Schottky diode
- Gunn diode

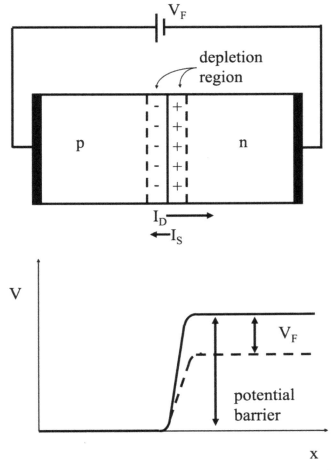

Figure 5.4. Forward biased diode junction. Top: diode junction. Bottom: potential barrier.

- Light emitting diode (LED)
- Photodiode.

5.2.1 Zener Diode

A Zener diode is a heavily doped p–n junction (Figure 5.11). It is designed with materials and a geometry appropriate to work with reverse bias in breakdown conditions. It has a very pronounced knee and allows one to control the voltage at the breakdown value (the current is defined by the circuit in which it is immersed).

It is used as a voltage regulator or to avoid overvoltages, due to spikes, for example. They are also excellent white-noise generators.

5.2.2 Schottky Diode

The Schottky diode is a doped metal–semiconductor junction (Figure 5.12). Its behavior is similar to that of the p–n junction but with a smaller threshold.

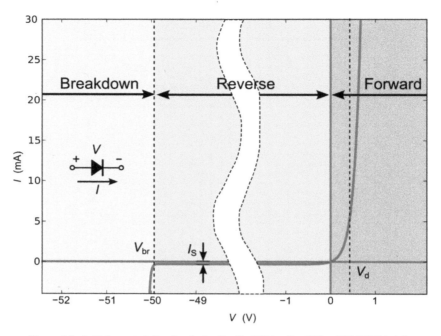

Figure 5.5. *I–V* characteristic of a diode. Credit: Wikipedia: Hldsc (CC BY-SA 4.0).

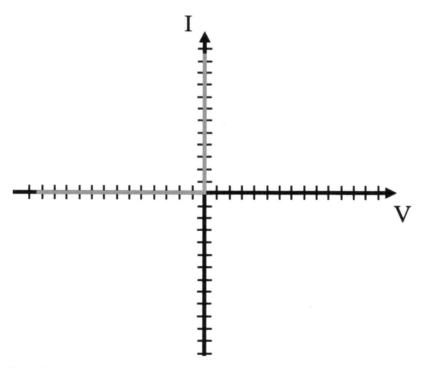

Figure 5.6. Diode as a bipolar component allowing current to pass only in one direction.

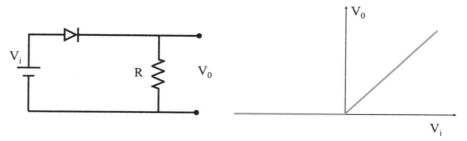

Figure 5.7. Left: diode in a circuit whose output voltage V_0 is zero if reversely polarized and a straight line if forward biased. Right: V_0 vs. input voltage V_i.

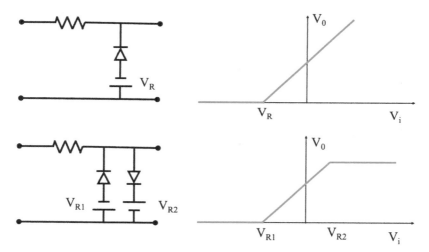

Figure 5.8. Diodes used as clippers.

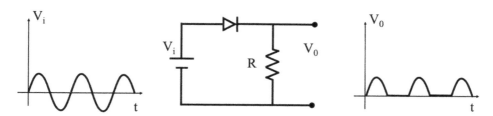

Figure 5.9. A diodes used as a rectifier. Left: input signal. Center: diode in a circuit used as a rectifier. Right: output signal.

The metal–semiconductor junction has a lower charge in the depletion layer so switching on–off is faster. Therefore, a Schottky diode has a higher speed than other diodes. That said, they have a lower reverse breakdown voltage (and a higher reverse leakage current). Often, they are manufactured with pre-installed series resistors. They are used as a quadratic component for radiometers.

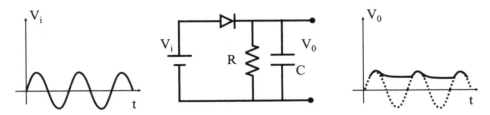

Figure 5.10. A diodes used as a rectifier. Left: input signal. Center: diode in a circuit used as a rectifier with a low-pass filter. Right: output signal.

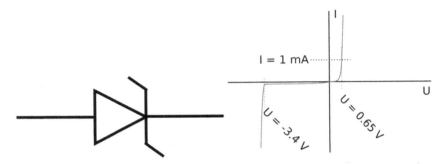

Figure 5.11. Left: a Zener diode and its symbol. Right: I–V curve of a Zener diode. Credit: Wikipedia: AlanM1 (CC BY-SA 4.0).

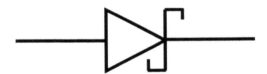

Figure 5.12. Schottky diode and its symbol.

5.2.3 Gunn (Tunnel) Diode

The Gunn diode (or transferred electron device, TED), is an n-doped semiconductor with two heavily doped outer zones and one more lightly doped interior (Figure 5.13). It is not exactly a diode because it can support currents in both directions.

Once polarized, a current is established. However, there is a voltage limit beyond which the thin layer inside increases the resistivity, and therefore the current decreases with increasing voltage: electrical resistance is thus $R < 0$. Current flows in the junction even if the potential energy of the carriers is below the voltage barrier. According to quantum mechanics, electrons may overcome a barrier higher than their energy. In this situation, oscillations occur in the Gunn diode in the microwave band due to the Gunn effect and the diode becomes a microwave source. When it is mounted in a cavity that serves as resonator, it allows one to vary the frequency and intensity of the radiation issued. The cavity sometimes mates to a waveguide and a horn in order to radiate power.

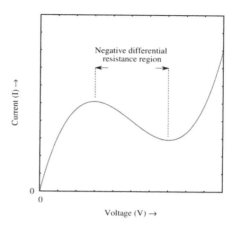

Figure 5.13. Left: Gunn diode with its integrating cavity. Right: *I–V* characteristic of a TED with its negative differential resistance region. Credit: Wikipedia: (CC BY-SA 3.0) https://it.wikipedia.org/wiki/Diodo_Gunn.

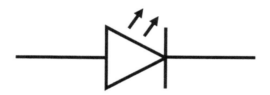

Figure 5.14. Symbol of an LED.

5.2.4 LED

A light emitting diode (LED) is a p–n junction of materials such that, when forward biased, it transforms the stored energy into electron–hole recombinations with the consequent emission of photons (Figure 5.14).

All diodes dissipate part of the energy of recombination in some form. Those made of germanium (Ge) or silicon (Si) dissipate it in heat, while those made of gallium arsenide emit photons in the visible. Different semiconductors emit different wavelengths in relation to the energy gap of the semiconductor.

5.2.5 Photodiode

A photodiode is an reversely biased p–n junction that converts photons into current (Figure 5.15).

For these applications it is customary to put an intrinsic semiconductor between zone p and zone n to create a PIN diode (which increases, among other things, the effective area presented to the photons). A photon with sufficient energy ionizes a bond (it is a photoelectric effect in the depletion area) and generates a gap electron pair that is carried away by the reverse electric field, and then a current is

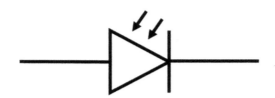

Figure 5.15. Symbol of a photodiode.

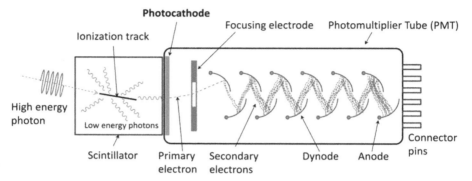

Figure 5.16. Image of a photomultiplier tube. Credit: Wikipedia: Qwerty123uiop (CC BY-SA 3.0).

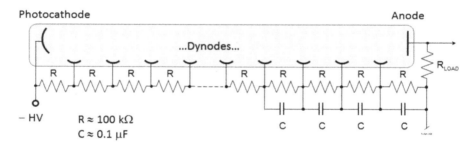

Figure 5.17. A typical photomultiplier circuit. Credit: Wikipedia: Qwerty123uiop (CC BY-SA 3.0).

established to compensate for the loss of electrons and holes. Even in the absence of light, there is a dark current to be controlled. A current is, however, present even if not polarized due to incident light (the same principle of operation as in photovoltaic cells). In Avalanche photodiodes, there is an extra layer that creates an avalanche photomultiplier effect. The ability to illuminate the sensitive part is therefore very important, so they are coupled with lenses and coated with anti-reflective materials.

5.3 Photomultipliers and Their Use in Astrophysics

A photomultiplier is another radiation detector that is particularly suitable for detecting IR–UV light (Figures 5.16 and 5.17). The basic principle is the photoelectric effect and electron multiplication in a vacuum glass tube. The sensor is directly next to a cathode that, by the photoelectric effect, emits an electron. This electron is conveyed and accelerated by a potential difference toward a series of potential electrodes (diodes) with a gradually higher potential (even up to 100 V) obtained with a series of resistive dividers. These in turn can emit electrons up to the anode. At the anode, electrons produce a detectable and digitally convertible current. Among the problems they have is the delay compared to the arrival and a noise level that is difficult to control, which is also amplified.

The following scientific cases can be studied through the use of photomultipliers:

- Cosmic rays: The study of cosmic rays involves the detection of particles with very high energies.
- Neutrinos: Neutrinos of extra solar system origin are fundamental to the study of γ-ray bursts, BLAZARs, and supernova remnants (as in the case of Alert IC170922), and are detectable thanks to the production of highly energetic muons.
- Dark matter: A measured excess of neutrinos could be evidence of WIMPs.

In almost all of these cases there is the need to reveal a highly energetic particle (typically a muon) that arrives from space or that is produced by a neutrino when it interacts with matter. In turn, these particles travel at relativistic speeds and produce Cherenkov light. It is essential to be able to determine the direction of travel and to reconstruct the origin of the relativistic particles in question (see Section 1.6).

In order to detect this light, one needs cheap and fast detectors that can work in difficult conditions: photodiodes (and photomultipliers). In addition, cosmic rays generate 10^6 muons in the atmosphere. For this reason, we need very powerful shielding systems (e.g., the Earth, and then the muons enter the detector from below). Large volumes and huge masses of media are needed to produce signals (e.g., the sea or the Antarctic ice pack). Directionality is needed to be able to see particles coming from below. The signal detection speed needs to be able to measure the Cherenkov cone and particle speed. Basically, what is needed is a 3D telescope of around a cubic kilometer in size. Cherenkov light detectors are also needed on the upper surface of the "neutrino telescope" to distinguish particles coming from above from those coming from below (so-called "anti-coincidence").

A prime example of such a "neutrino telescope" is IceCube[1] (described in Section 1.6). On the surface, Ice Top,[2] another bi-dimensional array, is used to identify particles that do not originate from the interaction between neutrinos and the ice.

IceCube has 5160 detectors placed between 1500 and 2500 m below the ice surface. They are distributed on 86 strings lowered into holes in the ice made through hot-water drilling. Detectors must have a time constant of less than 5 ns and a dynamic range of at least 200, must work down to $T_{min} = -55$ °C. The dissipated

---

[1] https://icecube.wisc.edu/
[2] https://icecube.wisc.edu/science/icetop/

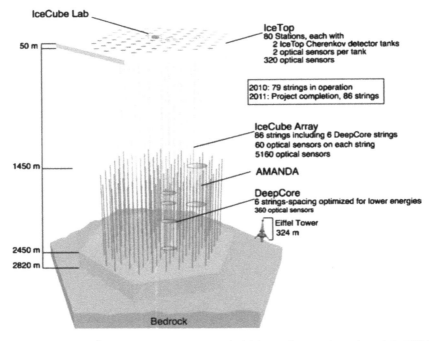

Figure 5.18. IceCube, a km³ neutrino telescope compared with its predecessor Amanda and the Eiffel Tower to give a sense of scale. Credit: Wikipedia: Nasa-verve—IceCube Science Team—Francis Halzen, Department of Physics, University of Wisconsin (CC BY 3.0).

power should not exceed $P = 5$ W. Furthermore, as the detectors are inaccessible once they are placed in the ice, they must have >90% reliability over 15 yr. Inserted into each borosilicate glass sphere are a photomultiplier, three LEDs (for calibration), and elaborate digital reading electronics that are based on field-programmable gate arrays (FPGAs).

5.4 Junction-gate Field-effect Transistors

A junction-gate field-effect transistor (JFET) is a tripolar doped semiconductor component. Actually, it is a diode because it has one junction, a p-doped channel with two electrodes, a depletion region and an n-doped with another electrode. However, like transistors, it has three connections. It consists of a doped "channel" p (or n) in which the current can flow. The three poles are the gate (G), the source (S) and the drain (D). A polarization, a potential difference, is applied at the source and drain ends. In the middle of the channel there is a zone doped in the opposite way n (or p), to which the third pole (an electrode) is applied. The junction p–n (or n–p) is therefore the between gate and the channel. If a voltage is applied to the lower (upper) gate and to both the source and drain gates, we have a reverse bias junction (a diode), and so the barrier increases.

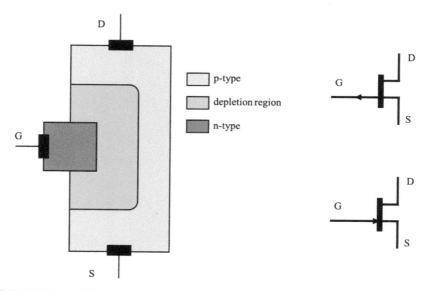

Figure 5.19. Left: the working principle of a JFET (see the text for details). Right: the symbol for a p-channel JFET (top) and an n-channel JFET (bottom).

Then, a depletion region is created in the channel. What is novel here is that the physical amplitude of the depletion region depends on the applied source–gate voltage. So, the width of the channel decreases and reduces the conductivity of the channel (and increases its resistance). Since the gate–channel junction is reverse biased, no current can flow there. The source–drain current will depend linearly on the source–gate voltage without the gate drawing current. We thus have created a mechanism that amplifies (maybe with amplification $A = 1$) an input voltage GS on the SD, but with an (almost) infinite input impedance.

JFETs are used to decouple faint signals that should be amplified (Figure 5.19). Typically, a semiconductor bolometer needs a JFET to be placed as close as possible to the bolometer to decouple from the subsequent low-noise amplifier.

5.5 Transistors

A transistor (its name arises from the union of "transfer" and "resistor") is a device made up of three oppositely doped parts (either npn or pnp), with the central part kept very thin (Figure 5.20).

It is a component that has three poles, an emitter (E), a base (B), and a collector (C). At the junction of the two regions, two depleting regions are created (like for a diode), and therefore two potential barriers are present (Figures 5.21 and 5.22).

Charge flow is due to diffusion across the thin junction that connects the two external parts. We have four different polarization combinations for the transistor relative to the forward or reverse bias of the individual junctions.

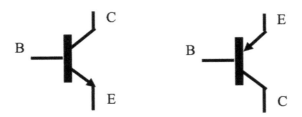

Figure 5.20. Left: an npn transistor. Right: a pnp transistor.

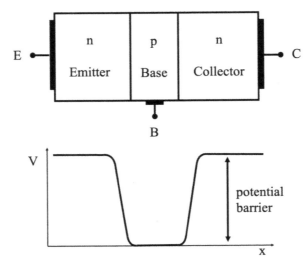

Figure 5.21. Top: an npn transistor. Bottom: the potential barrier present in an npn transistor.

In a typical active polarization:
- Base–emitter: forward biased (to decrease the potential barrier)
- Base–collector: reversed biased (to increase the potential barrier).

In this way the current flows from the emitter to the base, but since the latter is very thin, a small current passes through the base to the collector. The fact the base is thin means that any charge carriers injected into the base diffuse to the collector. We have that:

$$I_E = I_C + I_B \tag{5.2}$$

$$I_C = \alpha I_E \tag{5.3}$$

$$I_B = I_E(1 - \alpha) = \frac{I_C(1 - \alpha)}{\alpha} = \frac{I_C}{\beta} \tag{5.4}$$

with $\alpha \approx 1$ and $\beta \ll 1$.

In the above polarization scheme (an active biased polarization), the voltage between the base and emitter (V_{BE}) is positive at the base and negative at the emitter.

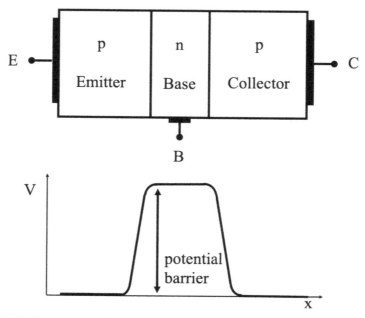

Figure 5.22. Top: a pnp transistor. Bottom: the potential barrier present in a pnp transistor.

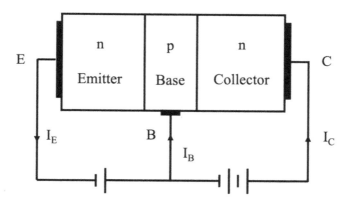

Figure 5.23. An npn transistor actively biased.

This is because, for an npn transistor, the base terminal is always positive with respect to the emitter (Figure 5.23). Also, the voltage supplied to the collector is positive with respect to the emitter (V_{CE}). So for an npn transistor to conduct, the collector is always more positive with respect to both the base and the emitter. The link between the input and output circuits is the main feature of the transistor because the amplifying properties of a transistor come from the consequent control that the base exerts upon the collector-to-emitter current. A large current (I_c) flows freely through the device between the collector and the emitter terminals when the transistor is switched on. This on state only happens when a small biasing current (I_b) is flowing into the base terminal of the transistor at the same time, allowing the base to act as a sort of current control input.

In general, a transistor must be properly biased to work correctly. This is done by using a network of resistors and voltage generators. A transistor in which the base is grounded, connected in forward bias, is a voltage amplifier of a current buffer (see Figure 5.24). A current applied to the emitter will be transferred almost entirely to the collector, so it can be seen as a current "buffer" (a component that keeps the same current independent of the load).

By choosing the collector resistance appropriately, a transistor in a common base configuration can be seen as a voltage amplifier. However, if we change the E–B voltage, a transistor becomes an E–C current generator controlled by an E–B voltage. Indeed, an E–C current can be modulated by the E–B voltage.

A transistor in which the emitter is grounded (a common emitter), connected in forward bias, is also an amplifier (Figure 5.25). A voltage V_{in} applied to the base, suitably chosen around the diffusion voltage of the E–B diode (the active region), produces a strong variation in the E–C current and thus in V_{out}. In fact, the ratio $I_B = \beta I_C$ is kept constant, and is $\beta \ll 1$ $I_C = I_B/\beta \gg I_B$.

A transistor has internal charge accumulation mechanisms that can be modeled with capacities that produce "high-pass" and "low-pass" responses. Typically, the high pass at low frequencies is linked to external capacities. Internal capacities, however, influence the response at high frequencies (Figure 5.26).

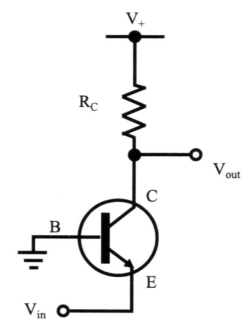

Figure 5.24. A common base transistor configuration.

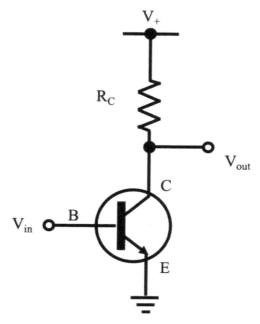

Figure 5.25. A common emitter transistor configuration as an amplifier.

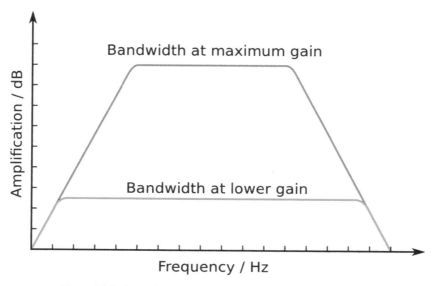

Figure 5.26. Spectral response of a transistor acting as an amplifier.

5.6 Operational Amplifier

If we now combine the functionality of two transistors, joined together, for example, with an emitter (which is a common configuration for an emitter amplifier), then we can amplify the difference between the voltages of the two bases (Figure 5.27).

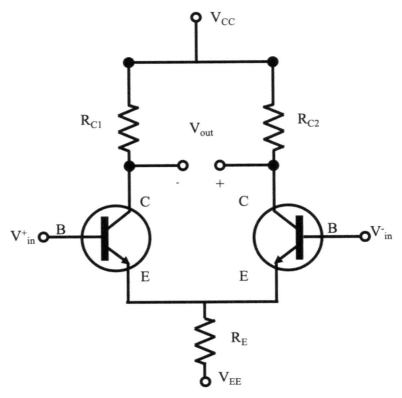

Figure 5.27. Two common emitter transistors forming a differential amplifier.

Here, the V^+_{in} signal provides a V_{out-} output with respect to the emitter. The V^-_{in} signal provides a V_{out+} output with respect to the emitter. So, the $V_{out+/-}$ signal will be proportional to the difference $V^+_{in} - V^-_{in}$.

A differential amplifier is a good choice, especially when any disturbance is present on the transmission line of a signal. We need to amplify a V_{in}, and if there is a disturbance that adds V_d noise on the transmission line, it will be amplified as well:

$$V_{out} = A(V_{in} + V_d). \tag{5.5}$$

However, if a differential amplifier is used, we have (Figure 5.28):

$$V_{out} = A[(V_{in+} + V_d) - (V_{in-} + V_d)] = A(V_{in+} - V_{in-}). \tag{5.6}$$

A set of transistors (e.g., pairs of transistors that form differential cascade amplifiers) forms an integrated circuit. Some of them are called operational amplifiers and are usually differential amplifiers characterized by a very high voltage gain (for example, up to 10^5–10^6), a high input, and a low resistance output.

By inverting the configuration and appropriately choosing the resistors in the circuit of a differential amplifier, a high gain can be obtained, and a feedback signal is inserted (at the negative pole) that counterbalances any gain fluctuations (Figure 5.29).

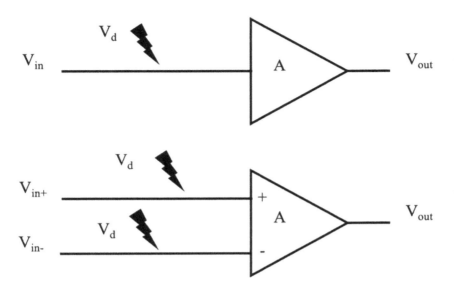

Figure 5.28. The difference between a single-ended amplifier (top) and a differential amplifier (bottom).

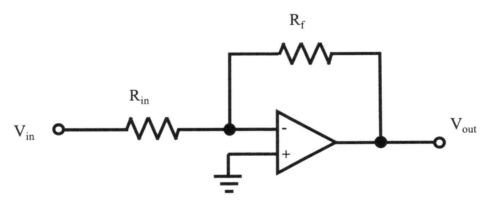

Figure 5.29. Differential amplifier in an inverted configuration.

If $R_f \gg R_{in}$ the current in the inverting input, the virtual mass (−), is:

$$\frac{V_{out}}{R_f} = -\frac{V_{in}}{R_{in}} \rightarrow \frac{V_{out}}{V_{in}} = -\frac{R_f}{R_{in}} \rightarrow V_{out} = -\frac{R_f}{R_{in}} V_{in} = A V_{in} \tag{5.7}$$

where the minus sign reflects the fact that the input signal is inserted at the inverting input. If R_f is large enough, the feedback current does not affect the input signal V_{in}. In addition, negative feedback has the effect of counterbalancing any changes in gain.

AAS | IOP Astronomy

Experimental Astrophysics

Elia Stefano Battistelli

Chapter 6

Detectors in Astronomy

Detectors used in astrophysics will be introduced in this chapter. The difference between quantum detectors, coherent detectors, and thermal detectors will be explained. The working principles of photoconductors, charge-coupled devices (CCDs), and other quantum detectors will also be detailed. Semiconductor and superconductor bolometers will be explained, including the bolometer equation, as well as X-ray microcalorimeters. Coherent receivers, classical radiometers, and the radiometer formula will be discussed, as will the working principles of heterodyne receivers and high-electron-mobility transistors.

6.1 Introduction

Among all of the carriers of astrophysical information, electromagnetic waves are the ones for which detector technology is the most developed. Among the different branches of astronomy, radio astronomy and optical astronomy are certainly the most advanced. The methods for detecting electromagnetic waves depend on their energy or frequency. An electromagnetic wave detector is a transducer that receives photons and, in most cases, produces an electrical signal that can be amplified and stored. We can summarily distinguish three different categories of detectors related to the energy of the detected photons:

- Coherent detectors (e.g., high-electron-mobility transistors, radiometers, antennas, etc.) detect low-energy (low-frequency) photons. They allow for both the amplitude and the phase of the incident electric field to be revealed by measuring the potential difference that arises at the ends of an antenna. They are intrinsically suitable for use as interferometers, but are less sensitive than thermal detectors (see below). These detectors work in the radio/microwave band.
- Thermal detectors (e.g., bolometers, Golay cells, etc.) detect photons with energies greater than those observed with coherent detectors. They do not react to the single photon, but exploit the integrated effect of a large number

of photons that produce a thermal effect (e.g., by heating the detector). Bolometers work in the millimeter, the submillimeter, and far-infrared frequencies.

- Quantum detectors (e.g., photodiodes, CCDs, and photographic plates) reveal high-energy photons. Every single photon produces an effect (e.g., the photo-emission of an electron). These detectors normally work from the mid-infrared to higher frequencies. For example, in a semiconductor, the incidence of a photon can produce a chemical change, vary the current in the crystal, or be amplified directly.

6.2 Coherent Detectors

While other detectors that measure radiation at higher frequencies are sensitive only to the amplitude of the incident signal (to the mean square value of the incident field), coherent detectors are able to reveal both the amplitude and the phase of an astrophysical signal. They are therefore intrinsically spectroscopic and sensitive to the polarization of the incident wave. In addition, because information about the phase is kept, signals from multiple detectors can be combined to make interference patterns. Coherent detectors are thus suitable as interferometers.

In radio astronomy the signals are revealed by "simple" antennas that convert the electromagnetic wave into a signal. We have three kinds of possible solutions. At low frequencies (~100 kHz), a dipole antenna in resonance with the electromagnetic wave is used. At higher frequencies (~MHz range), the same antenna has to be shielded to reject interference. A shielded antenna is thus used. At even higher frequencies (~GHz), radiation is collected using a horn antenna (Figure 6.1).

If the signal to be detected is at too high a frequency (for an amplifier and/or for an ADC), we can mix it with a locally generated "synthetic" signal (LO) and produce beats (see Figure 6.2). If this is processed with a square-law detector, it effectively produces a down conversion in the frequency of the signal. This is the basic working principle of heterodyne detectors.

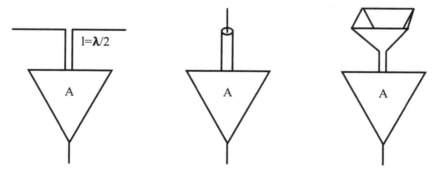

Figure 6.1. Left: for $\nu \sim 100$ kHz, a dipole antenna is connected directly to an amplifier. Center: for $\nu \sim 1$ MHz, a shielded dipole is connected to an amplifier. Right: for $\nu \sim 1$ GHz, a horn antenna is coupled to a waveguide and an appropriate transition to the amplifier.

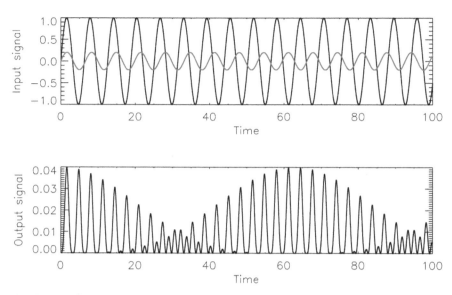

Figure 6.2. Top: two input signals into a heterodyne. Their frequencies are slightly different. Bottom: the input signals are processed by a square-law detector. The output has a component at low frequency (a beat phenomenon).

This kind of detector works in the Rayleigh–Jeans frequency range. At these frequencies, the power P emitted by a source is proportional to the thermodynamic temperature T. Thus, the characteristics of a radiometer (e.g., its noise or sensitivity) are all expressed in temperature units.

As mentioned above, coherent detectors measure both the amplitude and the phase of the incident electric field by measuring the potential difference that arises at the ends of an antenna. The Joule power dissipated in a resistor is:

$$P = \frac{V^2}{R} \rightarrow V \propto \sqrt{P}. \tag{6.1}$$

A device that conveys an electromagnetic wave in a transmission line is called an antenna. An antenna can emit and receive electromagnetic radiation. According to the reciprocity theorem, the properties and electrical characteristics of an antenna, such as the emission direction (the radiation pattern), the gain, and the resonant frequency, are the same whether it emits or receives. An antenna is an impedance adapter (it maximizes the transferred power) and can be a dipole or a reflector. This depends on the collecting area and directionality one needs to exploit:

- An antenna size $< \lambda$ is a so-called voltage element.
- An antenna size $\sim \lambda$ is a so-called resonant element.
- An antenna size $> \lambda$ is an optical element (as long as the network surface is adequate) that conveys the radiation of a funnel, i.e., a feedhorn.

In general, an optical element or a resonant element ends up with a probe antenna that measures the potential difference between the external chassis of the antenna

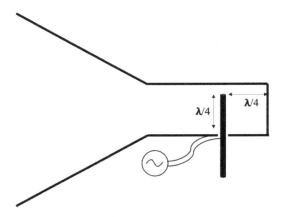

Figure 6.3. A probe antenna is an electromagnetic collecting element that detects radiation through a resonant element. The potential difference, in the case of a resonant element or an optical element, is measured between the external antenna structure and a probe that should be of size $\lambda/4$.

(the ground reference) and a resonant probe of size $\lambda/4$, positioned at $\lambda/4$ from the end point of the antenna (Figure 6.3).

6.2.1 Temperatures in Radio Astronomy

Noise Temperature

In radio astronomy power quantities are measured in terms of temperature that are additive. A resistance at a given temperature produces Johnson noise whose power is proportional to the physical temperature itself (see Chapter 8). From the noise power per unit bandwidth P_N, we can thus define the noise temperature T_N:

$$P = k_B T \rightarrow T_N = \frac{P_N}{k_B}. \tag{6.2}$$

Antenna Temperature

By analogy with the above definition, we can define the antenna temperature of a power P_A signal per unit bandwidth received from an antenna as:

$$T_A = \frac{P_A}{k_B}. \tag{6.3}$$

System Temperature

The signal output from the receiver will be due to the incident power P_A and the (noise) power of the detection system P_{sys}:

$$P_{tot} = P_A + P_{sys}. \tag{6.4}$$

We can then define a T_{sys} that quantifies the sensitivity of the receiver. Sometimes T_{sys} tends to include all sources of statistical noise that reduce sensitivity (e.g., receiver noise (rx), read-out amplification (RO), losses, antenna emission, spillover, atmospheric noise, etc.):

$$T_{sys} = \frac{P_{sys}}{k_B} \tag{6.5}$$

$$T_{sys} = T_{rx} + T_{RO} + T_{spillover} + T_{loss} (+ T_{atm} ...). \tag{6.6}$$

Brightness Temperature

In the Rayleigh–Jeans frequency range, the brightness temperature T_b represents the thermodynamic temperature of a blackbody (**BB**) that emits that brightness:

$$T_b = \frac{c^2}{2k_B\nu^2} BB(\nu, T) \tag{6.7}$$

where **BB** is the blackbody Planck function, c is the speed of light in a vacuum, k_B is the Boltzmann constant, and ν is the frequency.

If we are in the diffraction limit, the observed source is a blackbody, and there is a single polarization, then we have:

$$P_A = \frac{1}{2}A\Omega BB(\nu, T) = \frac{1}{2}\lambda^2 BB(\nu, T) = \frac{1}{2}\frac{c^2}{\nu^2}BB(\nu, T) \rightarrow T_A = T_b. \tag{6.8}$$

Given the definition of T_{sys} that quantifies the sensitivity of the receiver, in general it is necessary to take N independent measurements to detect a small signal in order to reduce the minimum detectable signal as a function of the square root of the N measurements. Considering the bandwidth of a non-spectroscopic (total power) receiver and the integration time, we have:

$$\Delta T |_{\Delta\nu=1Hz;t=1s} = T_{sys} \rightarrow \Delta T = \frac{T_{sys}}{\sqrt{N}} \xrightarrow{N=\Delta\nu\cdot t} \Delta T = \frac{T_{sys}}{\sqrt{\Delta\nu \cdot t}} \tag{6.9}$$

which is the so-called radiometer formula that establishes the sensitivity of a total-power radiometer for a given bandwidth and integration time.

An intrinsic limit to a radiometer is a quantum limit that arises from the Heisenberg uncertainty principle (Figure 6.4). A way of seeing it is that when we detect photons of frequency ν, we have a minimum uncertainty of +/− one photon in this measurement because we want to detect every single photon. This can be transformed into a noise temperature:

$$\Delta E_{min} = h\nu \rightarrow k_B T_{min} = h\nu \rightarrow T_{min} = \frac{h\nu}{k_B}. \tag{6.10}$$

This is an inviolable limit for all coherent receivers, and it increases with frequency.

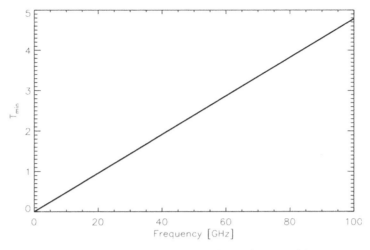

Figure 6.4. Radiometer noise quantum limit as a function of frequency.

6.2.2 Radiometers

A total-power radiometer measures a radio frequency (RF) within a certain bandwidth $\Delta\nu_{RF}$ determined by a filter. The detected frequency will thus be:

$$\nu_{RF} - \frac{\Delta\nu_{RF}}{2} < \nu < \nu_{RF} + \frac{\Delta\nu_{RF}}{2}. \tag{6.11}$$

When working with radiometers, it is advisable to have a detector that gives a signal proportional to the input power, so one needs a "square-law detector." We have that:

$$V_{in} \propto \sqrt{P} \rightarrow V_{out} \propto V_{in}{}^2 \rightarrow V_{out} \propto P. \tag{6.12}$$

Let G be the acquisition system gain. We have:

$$V_{out} = Gk_B T_A. \tag{6.13}$$

After the detection, the signal can be averaged over time and digitally converted. In summary, a radiometer is formed by (Figure 6.5):
- A $\Delta\nu_{RF}$ bandpass filter
- A square law detector
- A supplement/time average
- A reader or an ADC.

The effect of the bandpass filter centered around ν_{RF} and of a square-law detector is shown in the following formula, assuming a cosinusoidal input signal:

$$V_{in} \propto A\cos(2\pi\nu_{RF}t) \tag{6.14}$$

$$V_{out} \propto A^2\cos^2(2\pi\nu_{RF}t) = A^2[1 + \cos(4\pi\nu_{RF}t)]/2 \tag{6.15}$$

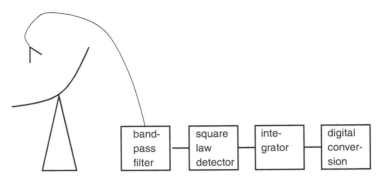

Figure 6.5. Scheme of a typical radiometer.

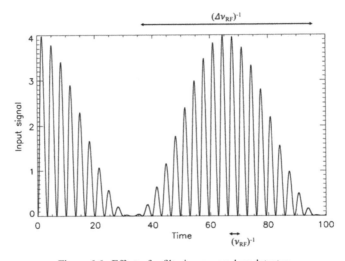

Figure 6.6. Effect of a filtering square-law detector.

so, we have a direct current (because the average is no longer zero) proportional to the input signal and a component at twice the frequency (which, however, is not needed). Figure 6.6 shows the effect of a square-law detector.

The integrator circuit integrates on a time t and allows for the determination of the astronomical signal (A) within the $\Delta\nu_{RF}$ band. A classic radio receiver must therefore have a quadratic or other type of non-linear device. Several devices studied so far can be used:

- A Schottky diode
- A junction-gate field-effect transistor (JFET)
- A cascade of transistor amplifiers
- A high-electron-mobility transistor (HEMT).

Schottky diode: The Schottky diode is a doped metal–semiconductor junction. It has a behavior similar to the p–n junction but with smaller voltages. Thanks to the small polarization voltage, a Schottky diode has a higher speed than do other diodes and can be used in radio astronomy. That said, they have a lower reverse breakdown voltage (a higher reverse leakage current). Often, they are manufactured with pre-installed series resistors. They are characterized by high noise (shot noise; see Chapter 8).

Junction-gate Field-effect Transistor (JFET): In a JFET, the physical amplitude of the depletion region depends on the applied source–gate voltage. So, the width of the channel decreases, which reduces the conductivity of the channel (increasing its resistance). Since the gate–channel junction is reverse biased, no current can flow there. The source–drain current will depend linearly on the source–gate voltage without the gate drawing current. In the JFET, we have created a mechanism that amplifies on the SD (possibly with $A = 1$) an input voltage GS, but with an (almost) infinite input impedance. The problem they have in radio astronomy is the high noise level they provide.

Transistor amplifiers cascade: We can have a series of n amplifiers with gain G_i. The total gain G is the product of all of the gains:

$$G = \prod_{i=1}^{n} G_i. \tag{6.16}$$

The total system temperature is therefore:

$$T_{\text{sys}} = T_{\text{rx}} + T_{s1} + \frac{1}{G_1}T_{s2} + \frac{1}{G_1 G_2}T_{s3} + \dots + +\frac{1}{G_1 G_2 \dots G_{n-1}}T_{sn}. \tag{6.17}$$

Cascade amplifiers must be used in such a way that the least noisy amplifier is used first. In general, they are limited to frequencies below a few GHz due to noise and the fact that they cannot be cooled down.

High-electron-mobility Transistor (HEMT): an HEMT is an evolution of a JFET. The materials used are mainly gallium arsenide (intrinsic) and the same doped n. An HEMT is therefore a heterojunction with a doped semiconductor facing a non- (or little-) doped one (Figure 6.7). As in all input and output field-effect transistors, they are largely decoupled due to the field effect.

The channel between the source and drain is actually the junction area between the substrate (S.I.-GaAs) and the doped area (n-AlGaAs). It is physically small, and electrons can flow in an area that is approximately two dimensional. Normally, most of the noise in transistors is related to electron scattering with the doped atoms of the lattice. High doping implies many electrons (high mobility), but also a lot of scattering (high noise). Due to the particular two-dimensional conformation of the motion of the electrons, scattering is largely reduced here, even when the mobility is high. In addition, an HEMT can be cooled down, therefore reducing all temperature-dependent noise. Thanks to the high mobility of electrons, HEMTs can be used at high frequencies (hundreds of GHz). "Commercial" HEMT amplifiers normally work at frequencies up to 10 GHz. HEMTs developed by research laboratories can go up to ν ~ 100 GHz and therefore can be used as radiometers at these frequencies. HEMTs are the most used receivers in radio astronomy at frequencies below 100 GHz.

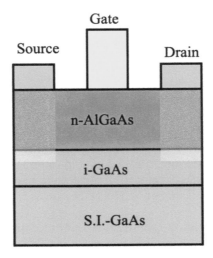

Figure 6.7. A high-electron-mobility transistor (HEMT).

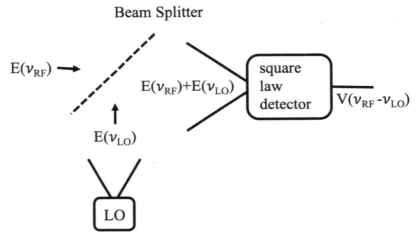

Figure 6.8. Optical (free space) configuration for a heterodyne receiver.

6.2.3 Heterodyne Receivers

At higher frequencies, or when one does not have an HEMT with an adequate performance, heterodyne receivers are used. This is also true in the case where one wants to minimize even more the auto-oscillations related to the fact that the frequency of the input signal to a receiver is equal to the frequency of the output signal.

The radio frequency (RF) signal is mixed with (added to) a signal synthetically produced by a local oscillator (LO) at a frequency similar but not identical to the RF frequency. Beats are produced at an intermediate frequency IF = |RF−LO|. A "down conversion" is carried out, and the IF frequency signal, which contains the astrophysical information, can be amplified by an HEMT or by a cascade of transistor amplifiers that work well at low frequencies (Figures 6.8 and 6.9).

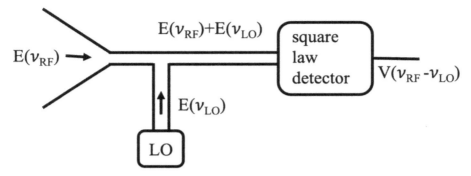

Figure 6.9. Waveguide configuration for a heterodyne receiver.

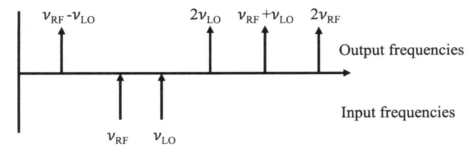

Figure 6.10. Input and output frequencies in a heterodyne receiver.

The key element of this detection chain is the mixer that receives the RF and LO signals by means of a beam splitter, waveguide, or a transmission line. Again, the mixer must have a quadratic response. A sinusoidal signal (both from the sky, ν_{RF}, and from the local oscillator, ν_{LO}) is processed by the mixer of gain G as follows. The output signal I is therefore (Figure 6.10):

$$I = G[E_{RF} \cdot \sin(2\pi\nu_{RF}t + \delta_{RF}) + E_{LO} \cdot \sin(2\pi\nu_{LO}t + \delta_{LO})]^2. \qquad (6.18)$$

This signal has several components:
- A DC component, $G(E_{RF}^2 + E_{LO}^2)$
- A second RF signal harmonic, $-\frac{1}{2}GE_{RF}^2 \cos(2 \cdot 2\pi\nu_{RF}t + 2 \cdot \delta_{RF})$
- A second LO signal harmonic, $-\frac{1}{2}GE_{LO}^2 \cos(2 \cdot 2\pi\nu_{LO}t + 2 \cdot \delta_{LO})$
- The sum frequency, $-GE_{RF}E_{LO} \cos[2\pi(\nu_{RF} + \nu_{LO})t + (\delta_{RF} + \delta_{LO})]$
- The difference frequency, IF, $GE_{RF}E_{LO} \cos[2\pi(\nu_{RF} - \nu_{LO})t + (\delta_{RF} - \delta_{LO})]$.

The signal of interest is the IF frequency, which can be adequately amplified and digitalized using a low-pass filter. RF signals above and below the LO frequency both are converted into an IF: the upper-side band (USB) and the lower-side band (LSB).

Due to the quadratic characteristic of the mixer, information about the sign of the intermediate frequency is lost:

$$\nu_{IF} = |\nu_{RF} - \nu_{LO}|. \tag{6.19}$$

This is not good if we want to do spectroscopy because the spectral information of the input signal is lost, the LSB mixes with the USB, and contaminations are created. A way to reduce this effect is to optically filter out either the USB or the LSB and operate in single side band conditions. That said, this effect could represent an advantage if one needs to make measurements in total power because the total band doubles (double side band), and the sensitivity drops by a factor of $\sqrt{2}$:

$$\Delta\nu = 2 \cdot \Delta\nu_{IF}. \tag{6.20}$$

Since the signal produced by the mixer is

$$I \propto G E_{RF} E_{LO} \cos[2\pi(\nu_{RF} - \nu_{LO})t + (\delta_{RF} - \delta_{LO})] \tag{6.21}$$

even a weak input signal E_{RF} can generate a strong output signal if the E_{LO} is powerful enough. The LO must then be powerful but also stable. An example one is a Gunn diode.

A well-performing device that is used as mixer in the millimeter and submillimeter bands is a junction of superconducting and insulator material. A superconducting material is a metal that, below a critical temperature T_C, has zero electrical resistance. In a superconductor, the electrons become a Bose condensate. In order for this to happen they must couple in pairs (a Cooper pair). Once superconducting, there is a "gap" of energy that must be supplied to the superconductor to "break" Cooper pairs and create quasi-particles, the electrons e^-. Two superconducting materials facing each other and that are insulated by a small layer of insulation comprise a device called a superconductor–insulator–superconductor (SIS) junction (Figure 6.11). If properly polarized, photon energies in the millimeter or submillimeter band can make a quasi-particle jump and create a current that passes through

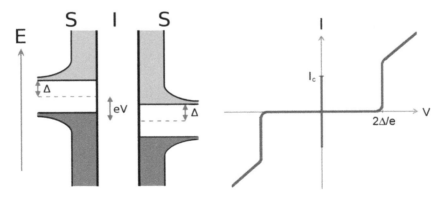

Figure 6.11. Left: energy states of an SIS junction. Credit: Wikipedia: NaniteVector: Zerodamage Tls60 (CC BY 3.0). Right: I–V curve of an SIS mixer. Credit: Wikipedia: (CC BY 3.0) https://en.wikipedia.org/wiki/Superconducting_tunnel_junction.

the insulation, producing a tunnel effect. The response is strongly non-linear and thus SIS junctions are excellent mixers in the millimeter and submillimeter regimes.

6.2.4 Gain Fluctuations and the Dicke Switch

The radiometer formula (Equation (6.9)) is an ideal formula; it tells us that, to reveal a small signal, we need to integrate for a long time: increase t. This is true only if the noise is of a statistical origin and there are no systematic effects. In reality, there are other effects that do not allow us to decrease the sensitivity ΔT at will. Among these effects we can list:

- The gain instability of radiometers and amplifiers
- The presence of the atmosphere
- Confusing noise

The gain G plays an important role here. Given a power, this produces an antenna temperature, as in the following relation:

$$P = Gk_B T_A \xrightarrow{\text{in case of no signal from sky}} P = Gk_B T_{\text{sys}}. \tag{6.22}$$

The contribution of the gain fluctuations (a false signal) must, unfortunately, be added to the noise. This is typically of type $1/f$ (see Chapter 8), so it indicates the maximum integration time for each observation:

$$\Delta P = \Delta G k_B T_{\text{sys}}. \tag{6.23}$$

Thus, we have an extra source of noise:

$$\Delta T_G = T_{\text{sys}} \left(\frac{\Delta G}{G} \right) \tag{6.24}$$

and so the total sensitivity is:

$$\Delta T_{\text{tot}}^2 = \Delta T_{\text{noise}}^2 + \Delta T_G^2 = T_{\text{sys}}^2 \left[\frac{1}{\Delta \nu \cdot t} + \left(\frac{\Delta G}{G} \right)^2 \right] \tag{6.25}$$

and:

$$\Delta T_{\text{tot}} = T_{\text{sys}} \left[\frac{1}{\Delta \nu \cdot t} + \left(\frac{\Delta G}{G} \right)^2 \right]^{0.5}. \tag{6.26}$$

To overcome the problem, we can use the Dicke switching method, which allows one to make differential measurements between two different "loads" and demodulate synchronously (Figure 6.12). The effect of the slower gain fluctuations of the modulation frequency are removed because they are common to the two branches of the observations. In fact, we have that:

$$P = Gk_B T_A \rightarrow Gk_B (T_A - T_{\text{ref}}) \tag{6.27}$$

and

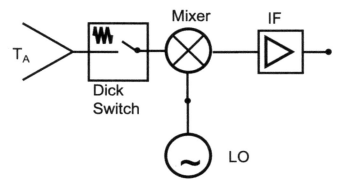

Figure 6.12. A Dicke switch radiometer: the signal is detected by the antenna and is sent to the mixer only half of the time. The other half of the time, the signal from a known and stable source is sent. After the mixer, the signal is amplified by a low-noise amplifier (LNA).

$$\Delta P = \Delta G k_{\mathrm{B}}(T_{\mathrm{A}} - T_{\mathrm{ref}}). \tag{6.28}$$

So, changes in gain are negligible. Of course, there is a price to pay. The disadvantage is that, if half of the integration time is passed on the load, then the minimum signal to be detected is increased:

$$\Delta T = \frac{\sqrt{2} \cdot T_{\mathrm{sys}}}{\sqrt{\Delta \nu \cdot t}}. \tag{6.29}$$

Atmospheric fluctuations can also be important at frequencies higher than ~5 GHz due to the presence of water vapor. To overcome this problem, we can use the Dicke switching method and make differential measurements between two different antennas (two different positions in the sky) and demodulate. Also in this case, we have a $\sqrt{2}$ loss in sensitivity.

Even when we can remove the effects of the receiver, the gain, the atmosphere, etc., there is a limit to the observations that is astrophysical in nature. At radio frequencies, when the angular resolution is a few minutes of arc, we have an important contribution from unresolved sources (such as radio galaxies). This is the so-called confusion noise. An empirical law dictates that it is a function of the frequency ν and of the angular resolution θ:

$$\sigma_{\mathrm{conf}} \,[\mathrm{mJy\ per\ beam}] \approx 0.2 \left(\frac{\nu}{1\,\mathrm{GHz}}\right)^{-0.7} \left(\frac{\theta}{\mathrm{arcmin}}\right)^2. \tag{6.30}$$

6.3 Thermal Detectors

Thermal detectors can reveal the integrated effect of many photons and can detect photons of lower energies than can quantum ones and higher energies than can coherent detectors. They take advantage of the strong variation of a physical characteristic with temperature, and since these characteristics depend weakly on the wavelength, they are intrinsically "broadband." Typically, thermal detectors are not sensitive to the polarization of the incident wave. However, they can be built in such a way to be sensitive to it, thanks to the optical/mechanical design. In order to select the bandwidth and polarization, they need optical elements in front of them, such as:

- Filters
- Quarter-wave plates
- Spectrometers
- Quasi-optical components

Among thermal detectors, we can list thermocouples, pyroelectric detectors, Golay cells, and bolometers, which in turn can be divided into semiconductor bolometers and superconductor bolometers, depending on the material they are made of. In some cases, the technology used in thermal detectors applies to X-ray detectors used as microcalorimeters, which, as opposed to thermal detectors, can detect the effect of a single photon.

6.3.1 Golay Cell

A Golay cell is a sensor that works at ambient temperature T_{amb} and is optimized for far-infrared radiation. It is a cell, a little room, that contains a gas (usually xenon) and has a transparent "window" to infrared radiation (Figure 6.13). A metallic film with low reflectivity acts as an infrared absorber and can be warmed up by incident radiation, spreading its heat to the gas in the cavity. In its expansion, the gas deflects a membrane on the rear wall of the cell. On the outside of the wall an optical system measures the deflection of a mirror that reflects the light coming from an LED.

A grid partially obscures the view of a photodiode that will reveal radiation proportional to the deflection and so proportional to the incoming infrared radiation. Usually Golay cells are used as laboratory tools. Reading problems occur in the presence of mechanical vibrations: they are very microphonic. The limit noise of a Golay cell is related to the Brownian motion of the gas particles (see Chapter 8). This

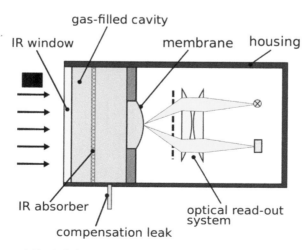

Figure 6.13. A Golay cell. Credit: Wikipedia: Ehab Ebeid (CC BY 3.0).

is an intrinsic limit related to the detector temperature. That said, it is not possible to cool down a Golay cell to cryogenic temperatures because the gas would liquefy. A Golay cell can be connected to a second cell with a weak thermal connection. In this way, slow changes in temperature, and even microphonics, can be greatly reduced.

6.3.2 Semiconductor Bolometers

A bolometer is a detector with an electrical resistance whose value varies greatly with temperature. The measure of the change of this resistance is a measurement of the heat coming from incoming radiation. It was invented in 1878 by the American

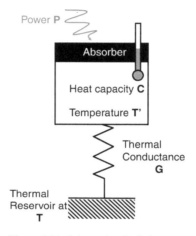

Figure 6.14. Schematic of a bolometer.

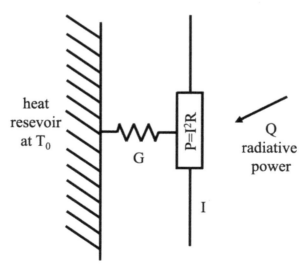

Figure 6.15. A semiconductor bolometer with dissipated power P that arises from the bias current I and radiative power Q.

astronomer Samuel Pierpont Langley. The most used bolometers so far are bolometers formed by suitably doped semiconductors. A bolometer is composed of an absorber that absorbs incoming radiation and a thermometer (a thermistor) that measures the change in temperature. This thermal mass, of head capacity C, is thermally connected with a "thermal reservoir," a cryogenic bath at temperature T_0, through a thermal conductivity G (Figure 6.14).

A current I flows and dissipates Joule power $P = I^2 R$. Furthermore, a radiative power Q clearly affects it: P and Q bring the bolometer to a temperature T higher than T_0 (Figure 6.15).

Neglecting the effect of the dissipated Joule power, the incident power Q on a bolometer increases its temperature of ΔT, depending on the thermal conductance G and the heat capacity C. The so-called bolometer equation holds:

$$Q = C\frac{d\Delta T}{dt} + G\Delta T. \tag{6.31}$$

If Q is constant (stationary conditions), we have:

$$C\frac{d\Delta T}{dt} = 0 \tag{6.32}$$

and so:

$$\Delta T = \frac{Q}{G}. \tag{6.33}$$

If now Q is brought to zero, we can solve the bolometer equation and we will have an exponentially decreasing temperature with a time constant τ:

$$\Delta T = \frac{Q}{G}e^{-\frac{t}{\tau}} \tag{6.34}$$

where:

$$\tau = \frac{C}{G}. \tag{6.35}$$

Now, we consider the effect of the dissipated power. The simplest reading circuit is formed by a battery that generates a constant potential difference. A load resistance $R_L \gg R$ is inserted in series with the bolometer. In this way, the current flowing in the bolometer is almost constant, and the voltage to be measured at the ends of it is:

$$V_{out} = A\left(V_{bias}\frac{R_{bol}}{R_L + R_{bol}}\right). \tag{6.36}$$

Any change in R_{bol} is then reflected in a change in the tension at its ends, which is then filtered (with a capacitor) and amplified (Figure 6.16).

Now, we can define the resistive parameter α that binds the temperature variations to the (relative) variations of resistance:

$$\alpha = \frac{1}{R}\frac{dR}{dT}. \tag{6.37}$$

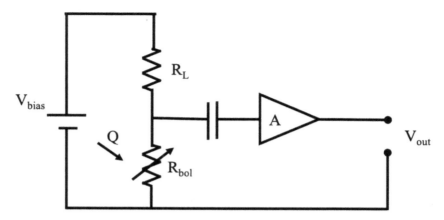

Figure 6.16. A read-out circuit for a semiconductor bolometer. The amplifier amplifies the signal at the ends of the bolometer after filtering high-frequency noise.

Bolometers can be built with different materials such as metals, semiconductors, and superconductors. We have that:
- Metals usually have $\alpha \sim 1$.
- Semiconductors have $\alpha \sim -10$.
- Superconductors have $\alpha \sim 1000$.

We can calculate the variation of the dissipated Joule power as a function of temperature. This is clearly a function of the α parameter and can be obtained as in the following:

$$\frac{dP}{dT} = \frac{d\left(\frac{V_{out}^2}{R_{bol}}\right)}{dT} = A^2 V_{bias}^2 \frac{d}{dT}\left[\frac{R_{bol}}{(R_{bol} + R_L)^2}\right] = A^2 V_{bias}^2 \left[\frac{R_L - R_{bol}}{(R_{bol} + R_L)^3}\right]\frac{dR_{bol}}{dT} \quad (6.38)$$

$$= \frac{A^2 V_{bias}^2}{(R_{bol} + R_L)^2} \cdot \frac{R_{bol}}{R_{bol} + R_L} \cdot \frac{R_L - R_{bol}}{R_{bol}} \cdot \frac{1}{R_{bol}} \cdot \frac{dR_{bol}}{dT} = P \cdot \frac{R_L - R_{bol}}{R_{bol} + R_L} \cdot \alpha.$$

This equation makes clear the dependence of the Joule power variation on the temperature and so on the incoming power. On one hand, we have that an incoming power Q increases the temperature of a bolometer, which then relaxes with its time constant. On the other hand, the same temperature variation produces a variation to the dissipated power P. Because α is negative for a semiconductor, a positive radiation power Q induces a negative dissipative power variation P, and so the bolometer is stable. This is an example of negative electrothermal feedback (ETF), which makes bolometers stable and actually usable. Later on, we will see an example of strong electrothermal feedback, which is stronger for superconductor bolometers.

A composite bolometer consists of two parts: an absorber and a thermometer (thermistor). In addition, a bolometer can be positioned in an integrating cavity that follows a concentrating horn. The absorber has the task of absorbing the electro-magnetic radiation and therefore must be made of absorbent material and must be

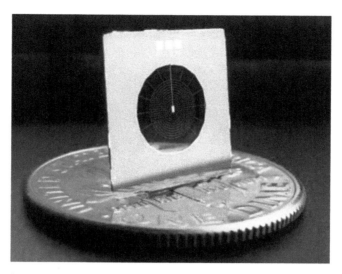

Figure 6.17. A spider-web bolometer built at NASA/JPL-Caltech. Credit: NASA/JPL-Caltech (Public Domain).

"large" compared to the wavelength (e.g., 1 mm^2). For this reason, semiconductor bolometers offer a large cross section to cosmic rays (especially in experiments positioned on high mountains, on stratospheric balloons, or on satellites), which generate strong pulses in the read-out signal depending on the bolometer time constant. One solution is to manufacture the absorber using a net of absorbing material: so-called spider-web bolometers (Figure 6.17). This reduces the cross section to cosmic-ray particles while maintaining the same cross section for millimeter radiation. Thermistors are made of a semiconductor material, typically doped germanium (Ge).

The bolometer invented by Langley was made of metal with $\alpha > 0$. However, the absolute value of α was pretty small and so its sensitivity was limited. In the 1960s, the first cryogenic bolometer was invented. Modern bolometers are made of doped Ge that, in order to have a uniform doping (without which we would experience changes in characteristics and increases in noise), is doped by "transmutation neutrons" from nuclear reactions that are then transformed into donor atoms or acceptors. This is neutron transmutation doped Ge, NTD-Ge. The thermistor is then glued onto the absorber and together they are suspended by means of low-conductivity material threads. Electrical contacts, welded by ultrasonic wire bonding, also serve as thermal contacts.

Semiconductor bolometers are intrinsically microphonic, meaning that when they vibrate, they pick up noise from environment fields. Their typical impedance is in fact on the order of a mega-Ohm, so any vibrations (which are always present in an astrophysical experiment) cause small currents that originate from the environment's electric and magnetic fields to translate into a large potential difference at their ends. In addition, modern bolometers are kept at cryogenic sub-K temperture while the amplifiers are at ambient temperature T_{amb}, which increases the length of the wires connecting a cryogenic bolometer to a T_{amb} amplifier. To overcome this problem, their output impedances must be decreased as quickly as possible. To do this, a cryogenic JFET is used that decouples the output signal from the bolometer from the signal to be amplified. The JFET must be placed as close as possible to the bolometer. JFETs do not work below 50 K, so they are kept at ~100 K in a Dewar that contains the bolometers.

6.3.3 Superconductor Bolometers

Just like semiconductor bolometers, superconductor bolometers are composed of an absorber and a thermistor. The thermistor, however, is a superconducting material placed at the critical temperature. They are called transition edge sensors, TESs. The stability of the temperature is fundamental; therefore, it is often "regulated" with feedback circuits.

Having $\alpha > 0$, the bias current cannot be constant, otherwise the electrothermal feedback would not work. We have to bias a superconductor bolometer with a constant voltage bias. Furthermore, $|\alpha_{super}| \gg |\alpha_{semi}|$ (Figure 6.18); therefore, the advantages of ETF are taken to the extreme. As a result of using superconductive bolometers, we have stability, a lower time constant, and noise reduction.

In order to ensure constant voltage across the TES, voltage losses along the electric connection from 300 K to 0.3 K (or less) must be minimized. Furthermore, changes in the resistance of the TES itself should not affect the voltage at its ends. It is then customary to send a constant current to the TES circuit and to place a shunt resistor with $R_{shunt} \ll R_{TES}$ in parallel with it. This ensures that the voltage across the TES is constant and that changes in resistance (i.e., due to incident power) translate into changes in current in the TES circuit.

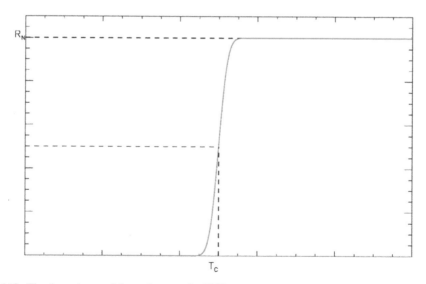

Figure 6.18. The dependence of the resistance of a TES on temperature. The critical temperature T_c is the temperature at which the TES transitions from its normal state, with an electrical resistance R_N, to its superconducting state, with zero electrical resistance.

The main difference between a TES and a semiconductor bolometer is linked to the radiation power, either when it is too small or too large. If it is too small, the ETF does not work properly, and the TES becomes a superconductor. If it is too large, the TES saturates and becomes normal.

In strong ETF conditions the current generated at the ends of a TES for an incoming power (its responsivity) acquires a very simple form. The responsivity R_e is defined as the change in current or voltage for a given incoming power. In the TES case we have a constant voltage, so we are interested in detecting the change in current. The derivative of the current with the incoming power P will then be:

$$R_e = \frac{dI}{dP} = \frac{d\left(\frac{P}{V_{\text{bias}}}\right)}{dP} = -\frac{1}{V_{\text{bias}}} \tag{6.39}$$

which is constant in the transition.

To read the current signal that passes through a TES, an inductance is inserted in the circuit of the TES itself. Variations of current are transformed into variations of magnetic flux that are detected by a magnetometer. The most sensitive magnetometers that exist are called superconducting quantum interference devices (SQUIDs; Figures 6.19 and 6.20). SQUIDs are superconducting devices and can be seen as rings of superconducting material with two Josephson junctions (shown as two crosses in Figure 6.19) kept on the normal/superconducting transition. They are very sensitive, have low noise, and are very fast. However, we should mention the fact that they have a strongly non-linear response.

As already mentioned, when a metal becomes superconducting, its resistance drops to zero. Another effect of superconductivity is the Meissner effect: in a closed

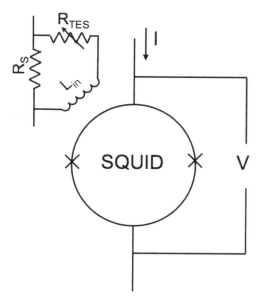

Figure 6.19. Signal from a TES inductively coupled (through L_{in}) with a SQUID.

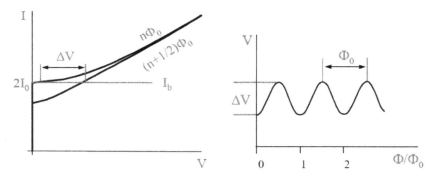

Figure 6.20. Left: maximum and minimum voltage read at a SQUID ends for different bias currents. Right: response of a SQUID. Voltage at its ends vs. the incoming magnetic flux for an ideal critical bias current. Credit: Wikipedia: Janolaf30 (Public Domain).

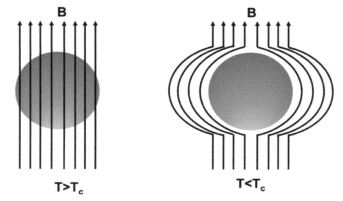

Figure 6.21. Left: a superconducting material at a temperature above the critical temperature inside a magnetic field. Right: the same material at a temperature below the critical temperature.

superconducting conductor, any applied magnetic field generates a current, which in turn generates an opposite magnetic field that cancels it. Surfaces or closed rings of superconducting materials "shield magnetic fields." The induced current can now be used to measure the applied magnetic field (Figure 6.20).

The current is produced by Cooper pairs (coupled electrons or bosons). The wave function that describes the current produced by the Cooper pairs is very "coherent" across the loop. Consequently, the current in a superconducting loop cannot assume all of the values because after a complete turn it must have the same initial phase. This leads to the quantization of the magnetic field, which can only assume values proportional to the quantum flux:

$$\Phi_0 = \frac{h}{2e} = 2.07 \times 10^{-15} \text{ Wb}. \tag{6.40}$$

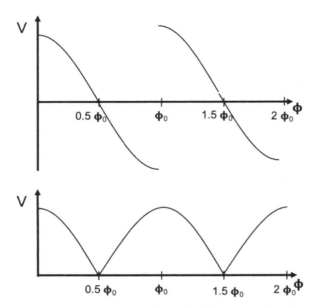

Figure 6.22. Top: voltage at the SQUID ends, subtracting flux up to the point at which it is more advantageous to add a period in the flux. Bottom: actual voltage read at the ends of the SQUID as a function of the applied magnetic flux, which results from the absolute value of the top curve.

When we apply a magnetic field, the current in the SQUID opposes the magnetic field. Assuming it is possible to read this current (this is the role of the Josephson junctions in the middle of the SQUID), in a closed loop, this changes the phase of the wave function until it becomes more advantageous, energetically, to add a quantum of magnetic flux instead of subtracting current. This can be seen in Figure 6.22.

In order to read the response of a SQUID it is necessary linearize its response, which can be done by trying to keep the reading on the same point. This is performed through a feedback circuit that sends an equal-and-opposite signal (a feedback) and stabilizes the SQUID on a point of its characteristic. A flux lock loop (FLL) circuit or a digital proportional, integral, differential (PID) loop calculates the right signal, an equal-and-opposite feedback (FB) signal, to make $V=V_{out}$ and also a variable called Error $= V-V_{out} = 0$. The FB signal is the astrophysical signal we want to detect and is proportional to the incoming radiation signal (Figure 6.23).

6.3.4 Kinetic Inductance Detectors

Another new kind of detector able to detect, among the rest, millimeter-wave signals is the kinetic inductance detector (KID). KIDs, however, are not thermal detectors. They can be considered quantum detectors (even though they detect the integrated effect of photons) with gap energies much lower than that of charge-coupled devices (CCDs). KIDs are composed of superconducting material. As already mentioned, below T_c electrons join to form Cooper pairs and produce "superconductivity."

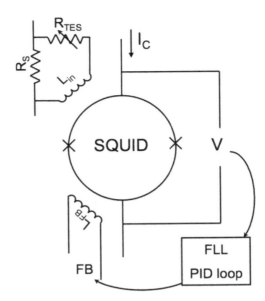

Figure 6.23. TES circuit coupled with a SQUID circuit, which is then read out by an ambient temperature feedback loop.

If a photon has enough energy to break a Cooper pair, then the pairs "break", which produces quasi-particles and induces a variation in the superconductor. The quantity that varies is the inertia opposed by the charge carriers to an AC voltage. Therefore, the kinetic inductance of the detector, which is inversely proportional to the density of Cooper pairs, varies. Thus, this is an inductance that, if combined with a capacitor, can form a microwave resonator. In fact, if it is coupled with a capacitor, we have a circuit with a variable resonance frequency. By monitoring the variation of the resonance frequency of the KID upon the heating of a photon, the signal is read. The frequencies involved are on the order of MHz–GHz.

6.4 Quantum Detectors

Quantum detectors such as photodiodes, CCDs, photographic plates, etc., detect high-energy photons. Every single photon produces an effect (e.g., the photo-emission of an electron). In order to detect photons in the visible we have various methods:

- The human eye, which has a logarithmic response and has a limited spectral efficiency
- A photographic plate, which has a greater spectral efficiency with respect to human eye
- A semiconductor detector, such as a CCD camera.

The quantum efficiency (the percentage of photons that are converted into the detected signal) of a photographic plate is similar to that of the human eye, but more extended in frequency. In addition, a photographic plate can integrate and can approach, in theory, as high a sensitivity as one desires. CCDs have a higher quantum efficiency and they also can integrate.

6.4.1 Photographic Plates

Historically, the first quantum detector is the photographic plate: a gel with silver bromide grains smaller than a micron in size is deposited on a glass substrate; when a high-energy photon interacts with it, it excites an atom and an electron passes from the valence band to the conduction band (Figure 6.24). Each electron is trapped by impurities or defects in the lattice, which acquires a negative charge. Silver ions are then attracted and form silver atoms, which in turn are additional impurities on which other atoms accumulate. At the end of the exposure, the silver atoms are found in the illuminated images so we have a negative film, in which areas of the plate illuminated by many photons appear darker than areas that are not. Then, the plate is *developed*, this process "amplifies" the Ag atoms, which are only a few units, until they form the latent negative image. Any rays transmitted into the substrate form halos; therefore, an anti-halo layer is positioned below it.

The chemical reaction responsible for the plate impression is described in the following:

$$AgBr + h\nu \rightarrow Ag^+ + Br + e^- \tag{6.41}$$

$$Ag^+ + e^- \rightarrow Ag. \tag{6.42}$$

Normally, the gels that are used are transparent at $\lambda > 235$ nm, so classic photographic plates cannot be used in the ultraviolet. That said, the minimum energy required to impress the plate is 2.82 eV, which corresponds to a wavelength $\lambda < 440$ nm, so photographic plates were not very useful in infrared astronomy. Once impressed, a photographic plate is "read" through a microdensitometer: the plate is illuminated and the transmitted radiation or its opacity is measured. The "opacity versus accumulated flux" response during exposure is strongly non-linear except in a specific zone.

Calibrations using photographic plates are (and were) complicated by the fact that they have to be carried out with different plates, which in principle have different response. The Palomar Observatory Sky Survey was conducted in the 1950s (POSS I) and 1980s (POSS II) and then digitized (scanned): http://archive.eso.org/dss/dss.

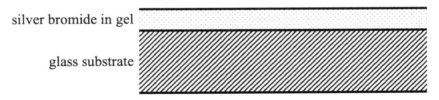

silver bromide in gel

glass substrate

Figure 6.24. Photographic plate.

6.4.2 Intrinsic and Extrinsic Photoconductors

Intrinsic photoconductors are simple semiconductor crystals, typically silicon or germanium: if the photon energy is greater than the energy gap between the conduction and valence bands of the semiconductor, then an electron–hole pair is created. The energy gap of silicon and germanium can be translated into wavelength as in the following:

$$E = h\nu \rightarrow \lambda = \frac{c}{\nu} = \frac{ch}{E} \sim \frac{1.24}{E(\text{eV})} \, \mu m \tag{6.43}$$

$$E_{\text{GAP}}^{\text{Si}} = 1.16 \, \text{eV} \rightarrow \lambda_{\min} \approx 1 \, \mu m$$
$$E_{\text{GAP}}^{\text{Ge}} = 0.75 \, \text{eV} \rightarrow \lambda_{\min} \approx 1.6 \, \mu m. \tag{6.44}$$

In single-pixel detectors, the semiconductor is "terminated" with metal contacts. A bias circuit generates an electric field and the photon flux generates a (photo-) current that can be amplified. Charge carriers can be created by incident radiation but also by thermal excitement: dark current.

Dark current must clearly be measured and minimized. The thermal excitation current decreases drastically, which cools the detector: most sensitive photodetectors must be cooled to 77 K (the boiling temperature for liquid nitrogen at 1 atm). The need to cool an intrinsic photoconductor is connected not only to the need to reduce dark current (which is normally measured and subtracted), but also to the need to reduce its fluctuations (Johnson noise and shot noise; see Chapters 8 and 9). When the intrinsic noise of a detector is lower than the intrinsic noise of the incident radiation (the optimal condition), we are in background-limited infrared photodetection (BLIP) conditions. Under these conditions, the efficiency with which the photons of electromagnetic radiation generate photocurrents is roughly independent of the wavelength (i.e., each photon "releases" an electron). The responsivity R_e is therefore proportional to the wavelength until the energy of the photon becomes greater than the energy of the gap:

$$R_e = \frac{i}{W} \propto \frac{1}{hc/\lambda} \propto \lambda. \tag{6.45}$$

The limiting (maximum) wavelength for an intrinsic photoconductor is linked to the gap energy of the semiconductor used. To work in the infrared, we need to decrease the gap energy. This is possible by "doping" the semiconductor. If impurities represented by "close" elements on the periodic table are inserted, an excess of electrons or conduction band holes are created, and the gap energy decreases. Though doped semiconductors can be used as photodetectors, there are some limits related to the doping density that cannot be exceeded. Also, doping creates low absorption coefficients for which large crystals (that are therefore sensitive to cosmic rays) are needed. Also, thermal noise from the dopant requires the temperature of the system to be lowered.

Examples of materials used as extrinsic photoconductors are listed in the following:

- The IRAS experiment (NASA) used arrays of ~60 extrinsic photoconductor detectors cooled to 2.6 K and made the first complete IR survey of the sky.
  - $\lambda = 4$ μm → Si: As
  - $\lambda = 25$ μm → Si: Sb
  - $\lambda = 60$ μm → Ge: Ga
  - $\lambda = 100$ μm → Ge: Ga.
- The Infrared Space Observatory (ISO; ESA) had a 60 cm telescope cooled to 1.7 K and enabled in-depth studies on more than 30,000 IR sources.
  - $\lambda = 2.5$ μm–5.5 μm → In: Sb
  - $\lambda = 5.5$ μm–17 μm → Si: Ga
  - $\lambda = 200$ μm → Ge: Ga
  - $\lambda = 100$ μm → Ge: Ga.
- Spitzer (NASA) is a satellite that is still in operation. It uses extrinsic photoconductive detectors in very large matrices.
  - $\lambda = 3.6$ μm–8 μm → Si: As (4×256^2 pixels)
  - $\lambda = 25$ μm → Si: As (4×128^2 pixels)
  - $\lambda = 24$ μm–160 μm → Ge: Ga ($128^2 + 32^2 + 20^2$ pixels).

6.4.3 Charge-coupled Devices (CCDs)

In an attempt to make a semiconductor memory device, Boyle and Smith of Bell Telephone invented a detector that they called a charge-coupled device (CCD) in 1969. CCDs are detectors that are inherently suitable for making images (like photographic plates) and can accumulate photo-produced charges (Figure 6.25).

Compared to photographic plates, CCDs have a larger sensitivity, they have a linear response, and they provide digital images (or at least electronic ones). CCDs are mosaics of intrinsic photoconductors. On each pixel (e.g., of Si, possibly slightly p-doped) a layer of silicon oxide is evaporated, and a layer of metal is evaporated above (a metal–oxide–semiconductor, MOS; Figure 6.26). The photoconductor is connected to ground while the metal is positively polarized. This polarization creates

Figure 6.25. Boyle and Smith winning the Nobel prize for physics in 2009. Credit: Wikipedia: © Prolineserver 2010, Wikipedia/Wikimedia Commons (CC BY-SA 3.0).

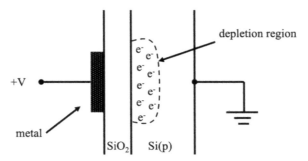

Figure 6.26. Metal–oxide–semiconductor (MOS), the working principle of a single pixel of a CCD.

a depletion zone in which the electrons produced by photons with $E > 1.12$ eV (for silicon) accumulate (Figure 6.26).

The electrons accumulate in an accumulation capacitor until their electric field counterbalances that of the polarization. Once the exposure is finished, the charges are carried sequentially along the pixels of a matrix up to a read-out field-effect transistor (FET). An intrinsic photoconductor generates a photocurrent when it is hit by electromagnetic radiation. The main source of noise is Johnson noise, and it is a current noise; therefore it is useful to increase its resistivity.

The MOS structure in a CCD is a capacitor and it provides the basic function of each pixel. The charge that can accumulate on one CCD pixel is given by the following expression:

$$Q_w = C_0(V_g - V_T) \tag{6.46}$$

where

$$C_0 = \varepsilon \frac{A}{x_0}. \tag{6.47}$$

Typical values are $V_g - V_T \sim 3$ V, $x_0 \sim 0.5$ μm, and $A \sim 25$ μm $\times$ 25 μm. We thus have $\sim 10^6$ electrons that can accumulate. Due to both noise and capacity issues, the resistivity of Si has to increase. In order to read these high-impedance detectors individually and quickly, they are coupled with MOS-FETs.

In order to be transparent to radiation (unlike real metals), the metal in a CCD is usually a highly doped Si layer . Highly doped Si, however, is not transparent to the UV, so at small wavelengths it is customary to illuminate the pixels from behind. The advantage of this is that it eliminates losses caused by the "metal" and oxide components (especially in the blue–UV), as well as by the various interfaces. The disadvantage is that the thicker the Si layer is, the greater the losses, so the Si layer must be very thin (~15 μm versus ~300 μm).

6.5 X-Ray Detectors

The bolometers that we have seen in this chapter can also be used to detect high-energy particles and X-ray photons. In this case they are called microcalorimeters. If the high-energy particle can release its energy into the absorber, its thermalization can be detected by the thermistor.

The same equation as for the bolometer applies:

$$Q = C\frac{d\Delta T}{dt} + G\Delta T \tag{6.48}$$

where:

$$Q = \begin{cases} 0 & \text{if } t < 0 \\ \Delta E/\Delta t & \text{if } 0 < t < \Delta t \, . \\ 0 & \text{if } t > \Delta t \end{cases} \tag{6.49}$$

It can be solved in the case of impulses, and then we will observe an exponential rise in temperature up to $\Delta T_{\max}$:

$$\Delta T = \frac{Q}{G}\left(1 - e^{-\frac{t}{\tau}}\right) \tag{6.50}$$

and

$$\Delta T_{\max} = \frac{\Delta E}{C} \, . \tag{6.51}$$

These microcalorimeters can detect even a few eVs, and the signal is proportional to ΔE. They are spectrometers, unlike CCD cameras, which can also be used as X-ray detectors. When used as (high-energy) photon detectors they need to be accompanied by similar detectors, such as anticoincidence detectors, that allow real X-ray photons to be distinguished from cosmic-ray particles. This is done by considering that photons completely release their energy in the absorber, while particles produce signals both on the absorber and on the anticoincidence absorber.

6.6 Detector Characteristics

6.6.1 Quantum Efficiency

Quantum efficiency is defined as the number of electrons produced in a detector divided by the number of incident photons (Figure 6.27). Quantum efficiency applies to quantum detectors and is the percentage of photons that are converted into a useful signal. For a photoconductor, quantum efficiency should be constant at high energies (short wavelengths) and suddenly fall to zero for energies lower than the gap energy. Other effects should also be considered. In order to improve quantum efficiency, one may need to increase the cross section offered by a detector to photons. For example, in a CCD, it is necessary to increase the thickness of the crystal. Doping has a similar effect. In addition, one may need to minimize the signal reflection between the vacuum and the substrate.

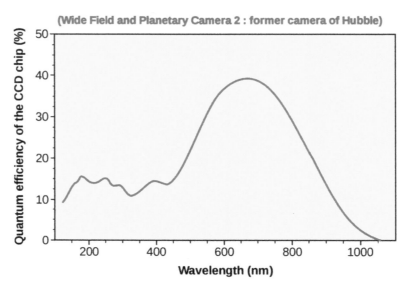

Figure 6.27. Quantum efficiency of a CCD on the Hubble Space Telescope. Credit: Wikipedia: Eric Bajart (CC BY-SA 3.0).

6.6.2 Linearity and Responsivity

A given incident power will produce an output signal. The direct proportionality between the input power (W) and the output signal (in voltage V or current intensity I) is called linearity. A linear detector has an added value because we can characterize its response through only one quantity, the responsivity. Responsivity is the measure of the input–output gain of the detectors and is expressed in $[R_e] = A/W$ or V/W and depends on the wavelength of the radiation. It is not necessarily a figure of merit for the detector (that is, one must consider also noise). Responsivity can have a response in frequency or wavelength. If, for instance, a CCD has a constant effect within a certain wavelength range, the responsivity will increase with wavelength up to the gap energy, at which the responsivity drops to zero.

6.6.3 Noise and Noise Equivalent Power (NEP)

During an astrophysical observation, the acquire data are usually affected by noise. Read-out noise is related to different sources, such as the ADC conversion and other electronic noise sources (Johnson noise, shot noise, etc.; see Chapter 8).

Any physical variable that fluctuates irregularly is said to have noise. In acoustics, noise is distinguished from "sounds", which instead have oscillatory behaviors. The "sounds" for an astrophysicist are the signals. Noise is one of the fundamental characteristics of an astrophysical observation. The first source of noise in an observation is the noise of the incident radiation that is detected. This can be related to:

- Intrinsic fluctuations of the signal to be detected (photon noise)
- The presence of the atmosphere

- The seeing
- The presence of other sources that are not of interest.

In addition, noise is also generated in a detector and/or in the reading/amplification chain. In large part, noise is linked to the thermal agitation of a system (but not only so), so it will be useful to learn cryogenic techniques (see Chapter 9). Clearly, noise degrades the measurement and therefore must be minimized and/or controlled. In any case it is necessary to measure noise in order to know the limits of each astrophysical observation.

Statistical fluctuations can be described by a Gaussian distribution centered around an average value and with a standard deviation. In order to improve the accuracy of a measurement affected by statistical noise, we have to increase the number of independent measurements: in this way, the uncertainty drops as the square root of the number of independent measurements. Since, for continuous measurements over time affected by uncertainty σ, the number of independent measurements n is proportional to the integration time t, the uncertainty is inversely proportional to the square root of time:

$$\sigma_{\text{final}} = \frac{\sigma}{\sqrt{n}} \propto \frac{\sigma}{\sqrt{t}}. \qquad (6.52)$$

If we want to compare different kinds of detectors, we have to define a quantity that describes their noise in terms of incoming power. The noise equivalent power (NEP) is defined as the power that in a second of integration produces a signal equal to the standard deviation of the detector noise: it is therefore the minimum signal detectable in a second of integration. If the standard deviation of this noise in volts or amperes is $\sqrt{\Delta S^2}$, then we have:

$$\text{NEP} = \frac{\sqrt{\Delta S^2}}{R_{\text{e}}}. \qquad (6.53)$$

6.6.4 Dynamic Range

The dynamic range of a detector is the ratio between the maximum detectable signal (detected with linearity) and the minimum detectable signal. Even if we change (e.g., digitally) the validity range of a digital detector, the dynamic range remains the same.

Semiconductor bolometers have a huge dynamic range (up to seven orders of magnitude). Superconducting bolometers instead suffer from saturation. Photographic plates had dynamic ranges greater than 100. CCD cameras can have dynamic ranges of four or five orders of magnitude. The importance of the dynamic range depends on the type of observations we make. In radio astronomy one needs a wide dynamic range, as we may detect sources of different fluxes (for instance) during calibration. In millimeter Comic Microwave Background (CMB) astronomy, we do not need a large dynamic range (especially when observed from space). At visible wavelengths, it depends on the scientific case. The minimum signal can be linked to the detector noise. Data compression techniques can alleviate dynamic range requirements.

6.6.5 Number of Pixels

One of the most attractive features of a receiver is the number of pixels. The number of pixels in an astronomical detector is a widely variable quantity. It can range from between "some" (for the first CCDs) to 1.4 gigapixels. CCDs with many pixels (together with an adequate focal length and pixel size) determine the field of view. This also has implications for the noise level and reading speed. Many pixels allow one to make maps of the sky and simultaneously improve sensitivity and decrease the integration time. The number of pixels that can be positioned at the focus of a telescope depends on the size of the focal plane (a diffraction-limited field of view, the physical size around the focus within which there are no distortions or else aberrations are controlled).

In the visible, we have very large arrays of detectors. They can be at the gigapixel level and up to 10^{10} pixels (Figure 6.28). In radio astronomy we typically have $N_{\text{pixel}} = 1$. CCD-like cameras are being built for use in the microwave range, thanks to the enormous development of multiplexing techniques that allow a large number of pixels (up to 10^4 pixels) to be read with a reduced number of wires entering into a camera, which is usually cryogenic (see Chapter 7).

6.6.6 Time Response

The minimum time interval within which a detector can "follow" a variation of the incoming signal is called the response time or time constant. A detector is always a low-pass filter and the cut-off frequency determines the time constant. We said that,

Figure 6.28. Rendering of the LSST camera to be installed at the Vera Rubin Observatory (Chile). The array's diameter is 64 cm. This mosaic will produce images that are over 3 gigapixels in size. Wikipedia: Todd Mason, Mason Productions Inc. / LSST Corporation (CC BY-SA 4.0).

in a CCD camera, the cross section (e.g., the dimensions) must be increased, but this also increases the response time. Sometimes it is not necessary to have a very fast response time. In this case there are advantages in terms of saved data and the operations carried out. In the case of quantum detectors, time constants can reach levels of 1 ns. Bolometers are slower (of order 1 ms), and the response time is proportional to the inverse of the responsivity. Superconductor bolometers, however, are faster than semiconductor bolometers.

6.6.7 Spectral Response

The spectral response is the frequency range of the radiation within which the detector can take measurements. A broad spectral response allows an observer to use a detector in different ways, for example, by using filters to select one band or another. Quantum detectors are typically limited to a wavelength above which they do not respond. For ground-based observations, the spectral response of a detector is clearly linked to the atmospheric transmission bands. Increasing the spectral response is often an advantage, but can lead to an increase in noise. Unlike the spectral response, the spectral band is the frequency range within which a detector is sensitive during an observation. The latter is at most equal to the former. Increasing the bandwidth means losing spectral resolution but increasing the signal.

For example, in radio astronomy one can average all of the frequencies detected by the detector and make observations in "total power." In optical astronomy, filters with predefined bands are used. Broadband filters are overlapping colored glasses. In order to have a higher spectral resolution, it is necessary to use interference filters (or spectrometers).

6.6.8 CCD Characteristics

A CCD has specific characteristics that we list in the following:
- Gain: The voltage (the number of electrons) necessary to produce an analog-to-digital unit (ADU) is called the gain, e^-/ADU (typical values are on the order of a few e^-/ADU). Gain values below 1 do not make sense, and even 1 e^-/ADU would lead to minimal signals immersed in noise, so it is not useful. Typical values are 5 e^-/ADU.
- Capacity: The number of electrons that saturate a pixel is called the capacity. Typically, it is 10^5 or 10^6 e^-;
- Dynamic range: The capacity divided by the number of bits of the ADC is called the dynamic range. For example, for a 16-bit ADC, there is a dynamic range of $10^5/16 = 6250$.
- Blooming: If the pixel capacity is exceeded due to an incorrect exposure, for example, we can experience an effect called blooming (the transfer to adjacent pixels). This can be mitigated. A similar phenomenon can occur during the transfer of charge for very bright sources if there is even a small leak between pixels.

- Dimension of the pixel: The pixel dimension is linked to the wavelength and to the telescope used. Typical values are about 10 μm. A greater pixel size has a higher capacity.
- Dark current: Dark current produces an offset that is dependent on the exposure time. It should be minimized by lowering the temperature, and typically it is fractions of e^- per pixel per second. At 290 K, 0.1 e^- pixel^{-1} s^{-1} produces 60 e^- pixel^{-1} in a 10 min exposure. At 230 K, 0.001 e^- pixel^{-1} s^{-1} produces a negligible effect.
- Gain changes: Not all pixels are the same, due to both intrinsic differences and the optics illuminating the CCD. Variations of up 10–15% from pixel to pixel can be observed. A way to mitigate the effect is to acquire a flat signal.
- Reading speed: Typical reading speeds vary from 50 kHz to 1 MHz. Considering the reading technique, this can cause an entire reading to be up to several seconds.

Experimental Astrophysics

Elia Stefano Battistelli

Chapter 7

Read-out Electronics

In this chapter we will introduce the basic concepts of read-out electronics for astronomical detectors. Analog versus digital electronics will be introduced; among other topics, we will discuss the CCD read-out system, low-noise amplifiers, SQUIDs read-out, the multiplexing technique, FPGA-based electronics, and microcontrollers.

7.1 Read-out Electronics

Astronomy has always given a strong impulse to the development of technologies and techniques, especially those related to analog and digital electronics. An example field of astronomy that gave a strong impulse to the development of read-out and control electronics is millimeter-wave astronomy. The management, polarization, and acquisition of data by millimeter detector arrays are very complicated, so there has been a need to develop specific read-out (RO) electronics tailored to an instrument.

Among the duties of read-out electronics are:

- The generation of the feedback signal for superconducting quantum interference devices (SQUIDs) and the cancellation of error
- The management of the multiplexing activity in a synchronized way
- The resonance search for kinetic inductance detector (KID) systems
- The search for polarizations and biases of thousands of SQUIDs and autotuning procedures
- Subsampling, filtering, and digital-to-analog (and vice versa) conversions.

Classic control electronics are no longer performing well: RO electronics with a high computing power and enormous computing speeds must be used. Today, we talk about electronics in the GHz range, managed by a field-programmable gate array (FPGA) that is programmed ad hoc. Examples of these RO electronics, developed in research institutes, are the University of British Columbia (UBC) Multi Channel Electronics (MCE[1]) or the Roach program developed at Berkeley.[2]

A FPGA is an integrated circuit whose functions are reprogrammable. The code with which this reprogramming is done is called firmware to indicate the position between hardware and software. This is a particular software written in machine language (and translated into a quasi-human language) that allows one to modify the hardware of the integrated circuit. It allows one to alter and configure the connections of millions of logical ports and in fact change the hardware of a processor by writing software. The fact that connections and logic gates are implemented in a single device makes FPGAs very fast devices that are gradually replacing most analog electronics. In addition, anyone in the electronics field can re-program an FPGA. This means an FPGA could potentially be included in a myriad of experiments. The FPGA manufacturing companies that dominate the market are Altera and Xilinx (Figure 7.1).

Given the complex operations of a read-out device, it should be tailored to a specific instrument. But it also must be versatile so a new one need not be developed every time there is a new experiment. Hybrid analog/digital RO electronics are needed with reprogrammable FPGAs. As a result, there has been a huge development in firmware. Electronics are needed that are:

- Modular
- Upgradeable
- Open source
- With standard (with industry) and modern communication protocols
- Low cost.

Also, low power consumption and a cooling system suitable for astrophysical experiments are needed for balloon-borne or satellite electronics and tests against cosmic rays (called "single event upset").

Figure 7.1. Altera (left) and Xilinx (right) FPGAs. Credit: (left) Wikipedia: Altera Corporation—Altera Corporation (CC BY 3.0), (right) Wikipedia: © Raimond Spekking / (CC BY-SA 4.0).

---

[1] https://e-mode.phas.ubc.ca/mcewiki/index.php/Main_Page
[2] https://casper.ssl.berkeley.edu/wiki/ROACH2

7.2 Multiplexing

The typical scheme for reading a transition edge sensor (TES) requires:

- A current bias for the TES, possibly shared between several TESs, which turns into voltage thanks to the presence of an $R_s \ll R_{TES}$ (in transition)
- A SQUID, or possibly a series of them (a SQUID series array, SSA[3])
- Warm (ambient temperature) electronics that operate the flux lock loop (FLL) circuit

TESs are so sensitive that, once they are photon noise–limited (i.e., limited only by the radiation hitting them), the only way to improve their performance is to increase the number of pixels. Sometimes the absorber is replaced with a planar antenna, followed by a microstrip line,[4] followed by a resistance that dissipates power into a TES (Figure 7.2).

However, increasing the number of pixels creates a problem from a cryogenic point of view. In fact, TESs or KIDs are typical cryogenic sub-Kelvin detectors. Arrays of thousands of pixels require several thousands of wires that must reach the sub-Kelvin temperatures of a cryostat. Their conductive thermal inputs would not allow the detectors to cool (Figure 7.3).

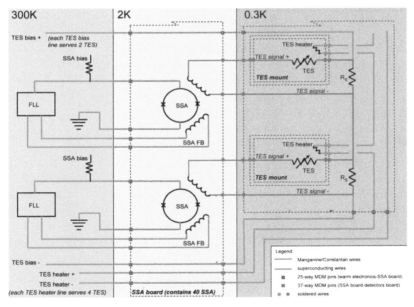

Figure 7.2. Typical read-out scheme of a TES array cooled down to 0.3 K. The TES is voltage biased through a shunt resistor R_s and its signal is passed to an SSA held at 4 K. The SSA non-linearity is managed with a feedback loop, an FFL, issued by ambient temperature read-out device.

---

[3] An SSA is a set of SQUIDS (of up to 100) put in series so that their gain is larger than that of a single one.
[4] A type of electrical transmission line in which a conductor is deposited into a dielectric used to transmit microwave frequency signals.

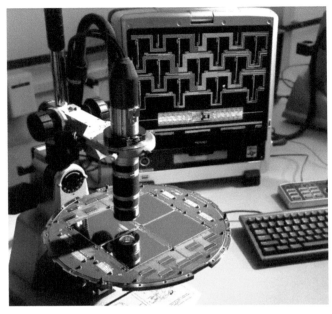

Figure 7.3. The BICEP2 detector array seen with a microscope. Credit: Wikipedia: Di NASA/JPL-Caltech.

With such a complicated and large array of detectors we need to reduce the amount of wiring that must reach the coldest stage of a cryostat. Multiplexing techniques can be used to do this. The detectors are grouped in blocks and all read together. To distinguish one from another, each is assigned a unique characteristic. Different techniques include:

- Time domain (or division) multiplexing, TDM
- Frequency division multiplexing, FDM
- Code division multiplexing, CDM.

TDM: In TDM, an array of detectors is divided into n rows and m columns (e.g., 32×32). Each TES is paired with a first-stage SQUID, SQUID1. The signals of all of the first stage SQUIDs in a column are added together and sent to another SQUID, the second-stage SQUID, SQUID2. Eventually, the SQUID2 signals are coupled to another SQUID stage with high gain, the SSA (not shown in Figure 7.4).

The SQUID1s of each row are connected together in series. The data acquisition is divided into n periods. At each period only one line of the array is on, and therefore in that period the signal read by SQUID2 is from only one row; one pixel for each SQUID2. This is possible because SQUIDs are fast and because, when they are not powered, they do not contribute any signal or noise (as they are super-conductors). In this sampling, the transmission of higher-level noise to the RO signal must be avoided, so a Nyquist inductor, an antialiasing low-pass filter, is needed.

FDM: If instead we insert N TESs, which are of variable resistances, into N resonant RLC circuits, they can resonate at different frequencies. We can send to the

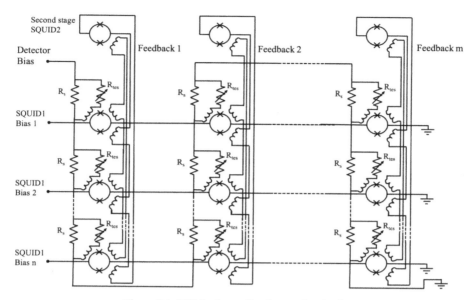

Figure 7.4. TDM scheme. See the text for details.

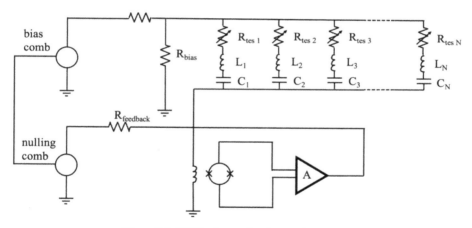

Figure 7.5. FDM scheme. See the text for details.

detector matrix a combination signal of N sinusoids that oscillate at the resonance frequencies and thus power all of the N TESs with only one line (Figure 7.5).

The read-out signal can then be demodulated, channel by channel, by monitoring the change in resistance of the TESs, which in turn changes the resonance frequency. This technique of multiplexing applies well to the read out of KIDs, where the incoming photons produce a change in the inductance of the detector and thus in the resonant frequency.

CDM: In CDM, the way to pair a TES with its SQUID is regulated by the Walsh matrix (see below). The columns of the matrix represent N successive periods of time.

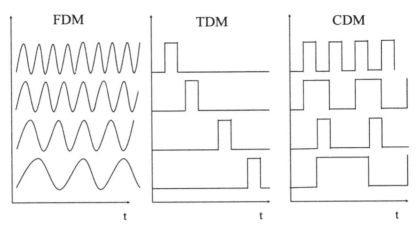

Figure 7.6. Time stamps of power sent to the SQUIDs for the three multiplexing schemes. Left: FDM. Center: TDM. Right: CDM.

Each of the rows that represent the different pixels to multiplex is "fed" following the time stamps dictated by the Walsh matrix:

$$W \equiv \begin{pmatrix} 1 & 1 & 1 & 1 \\ 1 & 1 & -1 & -1 \\ 1 & -1 & -1 & 1 \\ 1 & -1 & 1 & -1 \end{pmatrix}. \tag{7.1}$$

The Walsh matrix of TDM would be an identity matrix. CDM is an improvement on TDM, in that it allows one to keep each pixel "on" for a longer time compared to TDM.

The "−1" coupling between a TES and a SQUID can be achieved by inverting the coil of the coupling inductance between the two. Once the signals are acquired, we have to multiply the signal by the inverse of the W matrix: it is demodulated with the waveform of each row. The detector whose signal is read contributes all of its signals while the others will stay at zero.

To resume, in Figure 7.6 we show the power sent to the SQUIDs for the three different multiplexing techniques.

7.3 Read-out of a CCD

One of the fundamental characteristics of CCDs is that of being charge coupled. The pixels are isolated from each other and "collect" the photons. In order to read the CCDs, the content present in each pixel is "poured" into the adjacent one along a column and then it is collected through a shift-register parallel to the lines of the CCD.

Everything is then measured by a global collector synchronously. In reality three electrodes are used for each pixel and are polarized at different voltages. The voltages of the electrodes are changed cyclically so that the charges move from one electrode to another and then from one pixel to another: charge transfer. The charge transfer can have small losses. An efficiency of 99.9999% in modern CCDs is necessary since the transfer of charges, in CCDs of many pixels, may have to be carried out many times.

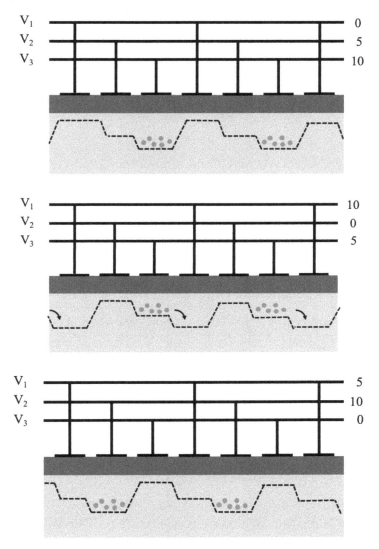

Figure 7.7. Read-out system of a CCD.

Once the electrons arrive at the serial register (which are also pixels of a CCD, but hidden), they are transferred one by one to an amplifier that amplifies and sends them to the ADC, which converts them to arbitrary densitometry units, ADUs (Figure 7.7).

7.4 Feedback Loop

A feedback (FB) loop is a control and feedback mechanism that makes changes to a system based on its current state and a desired state (Figure 7.8). It is used to regulate the functioning of a system and to compensate for any disturbances, adjusting the speed of correction. Feedback can be positive or negative. Positive feedback amplifies the force causing the error. In this case the system can be

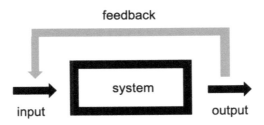

Figure 7.8. A feedback loop.

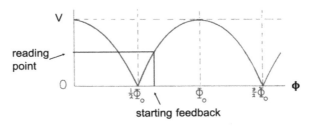

Figure 7.9. Non-linear SQUID response, shown as the voltage (V) output versus the input magnetic field Φ. The SQUID can be linearized, which keeps it at the same reading point with a feedback signal to the SQUID input.

unstable and diverge. Negative feedback opposes the force that causes the error. In this case the system converges (if the parameters are correctly chosen).

Examples of feedback loops in astrophysics are the adaptive optics that telescopes use to correct for atmospheric turbulence and the read-out system of a SQUID. In order to read and "linearize" the response of a SQUID it is necessary to try to keep the output on the same reading point (Figures 7.9 and 7.10). A feedback circuit is used to send an equal and opposite signal to the SQUID input that stabilizes it on a point of its characteristic. This circuit is an FLL circuit and can be an analog circuit or a digital proportional, integral, differential (PID) logic. A PID loop calculates the equal-and-opposite FB signal to make $V = V_{out}$ so that the Error $= V - V_{out}$ is kept at zero. The FB signal is the astrophysical signal of interest.

A PID controller is based on a simple proportional, integral, differential (PID) algorithm. It continuously calculates the difference between a variable to be controlled and a desired value (the set point) and applies a correction to the system through three factors P, I, and D (Figure 7.11).

The variable to be controlled is $r(t)$. The difference between $r(t)$ and the set point SP is calculated into the error variable $e(t)$. From the value of $e(t)$, the PID calculates the correction $u(t)$ using three parameters, K_P, K_I, and K_D, which multiply, respectively, the error variable (a proportional factor), the integral of $e(t)$ from the beginning of the experiment run (an integral factor), and the derivative of $e(t)$; (a differential factor). The variables are thus constructed as follows:

$$e(t) = r(t) - SP \qquad (7.2)$$

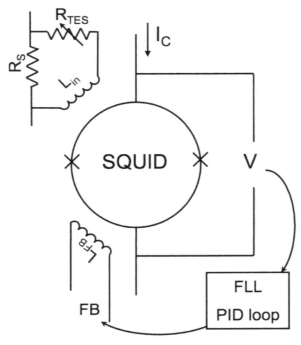

Figure 7.10. Coupling of a TES with a SQUID through an input coil. The output signal from the SQUID is processed by warm-temperature circuits, which sends to the SQUID a feedback signal that is equal and opposite to the input coil signal for the TES.

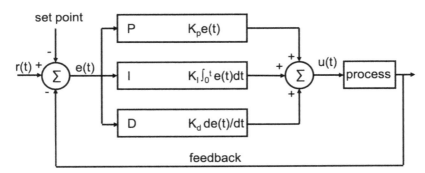

Figure 7.11. PID algorithm.

$$u(t) = K_p e(t) + K_i \int_0^t e(\tau)d\tau + K_d \frac{de(t)}{dt}. \tag{7.3}$$

As mentioned, the K coefficients are the P, I, and D coefficients that characterize the PID:

- K_P takes into account the distance of the signal from the desired SP and makes a proportional correction at this distance.

- K_I takes into account the previous values of the error and therefore has the ability to accumulate over time.
- K_D takes into account possible future trends because it calculates the speed of the approach to the SP and slows down if it is too fast.

The PID controller does not need to know the cause of the processes being observed, it merely measures them and applies a correction. For this reason, PID controllers have many applications. Just to mention a few:
- Control of the direction of a ship
- An FLL for a SQUID
- Temperature control (even for cryogenic systems)
- The movements of robots in factories
- Attitude checks in the flight of an airplane
- Power supplied by all power supplies
- Positioning the head of a magnetic hard disk
- Drone flights
- Adaptive optics in telescopes.

The operator that performs the parameter tuning (the choice of the three K parameters) is the one that must optimize the functioning of the PID in a specific case. A PID can be both analog and digital. Analog PIDs are made with integrative, differential, or proportional circuits so they have been useful in the past even though their parameters are not easily modifiable. By contrast, the ability to digitally modify a parameter of a circuit makes digital PIDs very popular. The most important parameter in a PID is K_P. The proportional term makes a correction proportional to how far away a variable is from the set point. Its gain is linked to the responsivity of the system: large gain corresponds to large responsivity, and vice versa.

However, if the term K_P is not well tuned (e.g., if it is too large), or if external factors intervene (noise, wind, etc.), then applying too much impetus in a correction causes the desired value to be exceeded (overshooting), and a reverse correction must be applied (Figure 7.12). If the tuning is not well done, the oscillation may even not attenuate, and the system would diverge. If K_p is too small (in particular for a digital PID), the corrective signal is small, and the system is unresponsive and less sensitive and it approaches the set point very slowly. In this case, the PID would not be able to correct quickly any disturbances, such as noise. Furthermore, for a digital PID, the corrective factor could be smaller than the minimum quantization level of the DAC that operates the control (Figure 7.13). This means that the $u(t)$ correction, once approximated by the quantization, is exactly zero so the system does not apply any correction to the system, even if there is still an error in the value: the steady state error.

One way to get around the problem of a steady state error is to add the integral term. An error, albeit small, will accumulate in the integral for a long period of time and will eventually apply a correction. The integral term takes into account the history of the system and, approaching the SP, accelerates the approach of the variable to be controlled to the SP and eliminates the steady state error. That said, due to the fact that it takes into account the history of the system, it is also slow and

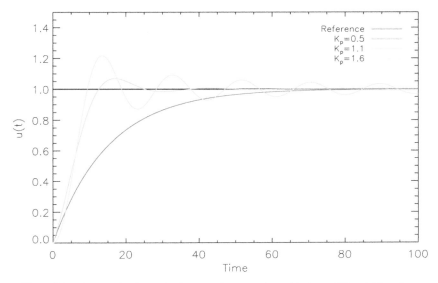

Figure 7.12. Approach of a variable to a reference value with three different K_P values.

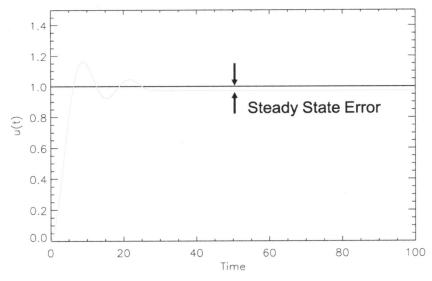

Figure 7.13. Occurrence of the steady state error for a digital PID.

could lead to inertia and overshooting, so it has to be fine-tuned. The addition of the differential term makes it possible to monitor the approach speed and, if it is too high, then the system can be slowed down. However, it is sensitive only to the last point of the thread, so it is sensitive to noise. In order to be used, the variable should probably be low-pass filtered.

Experimental Astrophysics

Elia Stefano Battistelli

Chapter 8

Noise and Its Origin

Noise and its physical origin are presented in this chapter. It will be introduced: Brownian motion, Johnson noise, temperature noise, shot noise, $1/f$ noise, fundamental noise, photon noise. Astrophysical applications will be presented.

8.1 Autocorrelation Function and Power Spectrum

A physical quantity that fluctuates in an irregular, non-deterministic way is known to be affected by noise (Figure 8.1). When the reasons for these fluctuations are manifold with multiple probability distributions and time constants, these statistical fluctuations can be described by a Gaussian distribution centered around an average value and with a standard deviation (the central limit theorem).

Noise has several origins that must be distinguished:

- Intrinsic to the radiation source
- External disturbances
- Detector + reading chain, of a fundamental or electronic type.

At least ideally, one can remove external disturbances (e.g., by shielding or analyzing the data in an appropriate way), so here we do not consider them. In order to improve the precision of a measurement affected by statistical noise, the number of measurements must be increased. As a consequence, the uncertainty drops as the square root of the number of independent measurements.

As for continuous measurements over time, the number of independent measurements is proportional to the integration time, and the uncertainty will be inversely proportional to the square root of time:

$$\sigma_{\text{final}} \propto \frac{\sigma_{\text{single}}}{\sqrt{n}} \propto \frac{1}{\sqrt{t}}. \tag{8.1}$$

Noise must be characterized, known, and studied. For example, it is necessary to quantify its level and study its probability distribution. As a stochastic process, noise

doi:10.1088/2514-3433/ac0ce4ch8

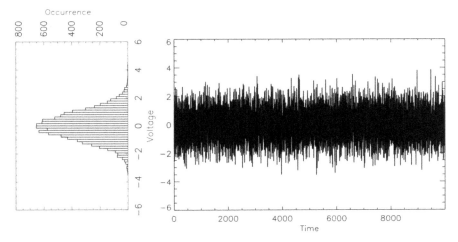

Figure 8.1. Signal affected by noise. Left: a histogram of noise. It is centered around zero. Right: a signal as a function of time.

cannot be predetermined at any moment in a deterministic way, so the treatment involves the use of average variables.

8.1.1 Average Quantities: the Autocorrelation Function

We hereby define some quantities that will enable us to describe and characterize noise.

- Average value:

$$\langle x(t) \rangle = \lim_{T \to \infty} \frac{1}{T} \int_0^T x(t) dt \qquad (8.2)$$

- Mean square value:

$$\langle x^2(t) \rangle = \lim_{T \to \infty} \frac{1}{T} \int_0^T x^2(t) dt \qquad (8.3)$$

Note that noise has a mean of zero, but the sum of signal and noise does not. Also, one should be careful if one wants to calculate the root-mean-square (rms) value of a non-zero-mean signal.

- Autocorrelation Function:

$$R(\tau) = \langle x(t) \cdot x(t + \tau) \rangle = \lim_{T \to \infty} \frac{1}{T} \int_0^T x(t) \cdot x(t + \tau) dt \qquad (8.4)$$

The autocorrelation function is the average of the signal multiplied by the signal translated by τ (Figures 8.2 and 8.3). It is thus a function of τ and tells us how quickly a signal evolves. $R(0)$ is the mean square value and, for a non-periodic

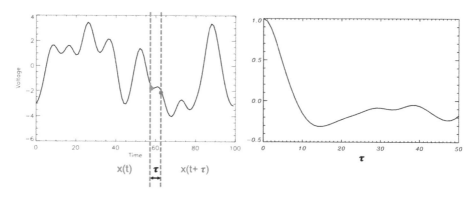

Figure 8.2. Left: a slowly varying signal. Right: its autocorrelation function.

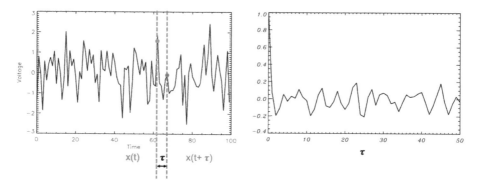

Figure 8.3. Left: a rapidly varying signal. Right: its autocorrelation function.

signal, it decreases as τ increases. For periodic signals it has maxima for τ multiples of the period of the function.

8.1.2 The Power Spectrum

Given a signal $x(t)$, we can derive the signal $x_T(t)$ by truncating $x(t)$ on the period T. We have then:

- $x_T(t) = x(t)$ for $t < T$
- $x_T(t) = 0$ for $t > T$.

We can make the Fourier transform $g_T(\omega)$ of $x_T(t)$. The "power" spectrum of $x(t)$; (note, it is not actually a power) is defined as the limit for T tending to infinity of the square of the $g_T(\omega)$ divided by T:

$$S(\omega) = \lim_{T \to \infty} \frac{g_T^2(\omega)}{T} = \lim_{T \to \infty} \left\{ \frac{1}{T} \left[\int_0^T x(t) \cdot e^{-i\omega t} dt \right]^2 \right\}. \tag{8.5}$$

$S(\omega)$ is a positive function that tells us, at a certain frequency (or pulsation, $\omega = 2\pi\nu$), how much "power" per band unit is carried by the signal. It is a spectral density and typically has units of $V^2 \, Hz^{-1}$.

The Parseval identity, or Plancherel's theorem, for "energy" (when it is finished should be replaced with finite) is valid and can be formulated as follows:

$$\lim_{\omega_0 \to \infty} \int_0^{\omega_0} g(\omega)^2 d\omega = \lim_{T \to \infty} \int_0^T x(t)^2 dt \tag{8.6}$$

and thus

$$\lim_{\omega_0 \to \infty} \int_0^{\omega_0} S(\omega) d\omega = \lim_{\omega_0 \to \infty} \int_0^{\omega_0} \left[\lim_{T \to \infty} \frac{g_T^2(\omega)}{T} \right] d\omega = \lim_{T \to \infty} \frac{1}{T} \left[\lim_{\omega_0 \to \infty} \int_0^{\omega_0} g_T^2(\omega) d\omega \right]$$

$$= \lim_{T \to \infty} \frac{1}{T} \int_0^T x^2(t) dt = \langle x^2(t) \rangle. \tag{8.7}$$

This is a sort of energy conservation theorem that states that the energy calculated in frequency space by integrating over all of the frequencies is the same as that calculated in time space by averaging over time.

8.1.3 The Wiener–Khinchin Theorem

The Wiener–Khinchin theorem states that the power spectrum of a signal is the Fourier transform of the autocorrelation function. If we now consider the definition of a power spectrum, we have:

$$S(\omega) = \lim_{T \to \infty} \left\{ \frac{1}{T} \left[\int_0^T x(t) \cdot e^{-i\omega t} dt \right]^2 \right\}$$

$$= \lim_{T \to \infty} \frac{1}{T} \int_0^T x(t) \cdot e^{+i\omega t} dt \cdot \int_0^T x(t) \cdot e^{-i\omega t} dt \tag{8.8}$$

$$= \lim_{T \to \infty} \frac{1}{T} \int_0^T \int_0^T x(t_1) \cdot e^{+i\omega t_1} \cdot x(t_2) \cdot e^{-i\omega t_2} dt_1 dt_2.$$

Now we can change variables: t_1 into t and t_2 into $t + \tau$. We have:

$$S(\omega) = \lim_{T \to \infty} \frac{1}{T} \int_0^T \int_0^T x(t) \cdot e^{+i\omega t} \cdot x(t + \tau) \cdot e^{-i\omega t} \cdot e^{-i\omega\tau} dt d\tau$$

$$= \int_0^\infty \left[\lim_{T \to \infty} \frac{1}{T} \int_0^T x(t) \cdot x(t + \tau) \cdot e^{+i\omega t} \cdot e^{-i\omega t} dt \right] \cdot e^{-i\omega\tau} d\tau \tag{8.9}$$

$$= \int_0^\infty \left[\lim_{T \to \infty} \frac{1}{T} \int_0^T x(t) \cdot x(t + \tau) dt \right] \cdot e^{-i\omega\tau} d\tau$$

$$S(\omega) = \int_0^\infty R(\tau) e^{-i\omega\tau} d\tau. \tag{8.10}$$

This demonstrates that the power spectrum is indeed the Fourier transform of the autocorrelation function. The measurement of the autocorrelation function is a powerful method to calculate the power spectrum of the signal.

8.2 Brownian Noise

Suppose we want to describe the motion of a particle in a viscous fluid (subject to viscous friction). A particle of mass m, velocity $v(t)$, and position $x(t)$ in general will undergo the average effect of the viscous force opposite to the velocity, and the instantaneous force F_{inst}, which is a function of the time t due to the impacts with individual fluid molecules. The viscous force is a force that depends on particle velocity and a constant B (the mobility of the body in the fluid). We have (Figure 8.4):

$$ma(t) = m\frac{dv(t)}{dt} = F_{inst}(t) - \frac{v(t)}{B}. \tag{8.11}$$

We can make the following assumptions for $F_{inst}(t)$; (the averages are ensemble averages of subsystems with the same properties):

- Isotropy: The force has no privileged directions, therefore:

$$\langle F_{inst}(t) \rangle = 0. \tag{8.12}$$

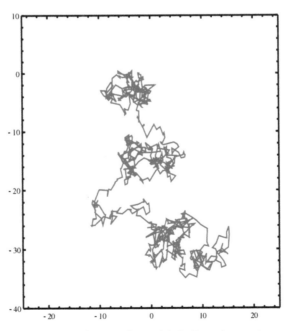

Figure 8.4. Trajectory of a particle in Brownian motion.

- Uncorrelation: The force originates from rapid impacts that are not correlated within each other; therefore, the force does not correlate with itself in subsequent instants (for large enough Δt). S quantifies strength!

$$F_{\text{inst}}(t) \cdot F_{\text{inst}}(t + \tau) = S\delta(t).$$ (8.13)

- Gaussian: The force is the result of a very large number of interactions; therefore, by the central limit theorem, its distribution will be Gaussian.
- The interactions produce a non-negligible effect on impact:

$$\langle F_{\text{inst}}(t)^2 \rangle \neq 0 = S.$$ (8.14)

From the mobility we can define the relaxation time $\tau_1 = mB$. Thus:

$$\frac{dv(t)}{dt} = \frac{F_{\text{inst}}(t)}{m} - \frac{v(t)}{mB} \rightarrow \frac{dv(t)}{dt} = -\frac{v(t)}{\tau_1} + \frac{F_{\text{inst}}(t)}{m}.$$ (8.15)

This differential equation can be solved by integration and has the following solution (assuming $v(0) = v_0$):

$$v(t) = v_0 e^{-\frac{t}{\tau_1}} + \int_0^t e^{-\frac{t-t'}{\tau_1}} \frac{F_{\text{inst}}(t')}{m} dt'.$$ (8.16)

From this equation of $v(t)$, we can derive the autocorrelation function of $v(t)$; (for $t \gg \tau$)

$$R(\tau) = \langle v(t) \cdot v(t + \tau) \rangle = \frac{\tau_1}{2} \cdot \frac{S}{m^2} e^{-\frac{\tau}{\tau_1}}$$ (8.17)

where $F_{\text{inst}}(t)$ is the instantaneous force caused by collisions with the molecules for which the above isotropy, correlation, and Gaussian properties apply.

We note that the mean square value (i.e., $\tau = 0$), for very large times is:

$$R(0) = \langle v(t)^2 \rangle = \frac{\tau_1}{2} \cdot \frac{S}{m^2}.$$ (8.18)

The initial velocity (v_0) is damped and the particles reach equilibrium with the fluid. The particle therefore has an average quadratic velocity produced by impacts. For long periods of time, statistical mechanics are in effect. We have, from the equipartition principle, that:

$$\frac{1}{2}m\langle v(t)^2 \rangle = \frac{1}{2}k_B T.$$ (8.19)

From which:

$$\frac{\tau_1}{2} \cdot \frac{S}{m^2} = \frac{k_B T}{m}$$ (8.20)

and so:

$$S = 2m\frac{k_{\mathrm{B}}T}{\tau_1}. \tag{8.21}$$

Then, the autocorrelation function is:

$$R(\tau) = \langle v(t) \cdot v(t + \tau)\rangle = \frac{\tau_1}{2} \cdot \frac{S}{m^2}e^{-\frac{\tau}{\tau_1}} = \frac{k_{\mathrm{B}}T}{m}e^{-\frac{\tau}{\tau_1}}. \tag{8.22}$$

The power spectrum, using the Wiener–Khinchin theorem, is:

$$S(\omega) = \int_{-\infty}^{\infty} \frac{k_{\mathrm{B}}T}{m}e^{-\frac{\tau}{\tau_1}} \cdot e^{-i\omega t}d\tau = \frac{k_{\mathrm{B}}T}{m} \int_{-\infty}^{\infty} e^{-\frac{1+i\omega\tau}{\tau_1}\tau}d\tau = \frac{4k_{\mathrm{B}}T\tau_1}{m(1 + \omega^2\tau_1^2)} \tag{8.23}$$

and so, if $\omega \ll 2\pi/\tau_1$, we have

$$S(\omega) = 4k_{\mathrm{B}}T\frac{\tau_1}{m}. \tag{8.24}$$

8.3 Kinds of Noise

Here we list the different kinds of noise we will discuss in this section:

- Johnson noise: This is caused by fluctuations (agitations) of a conductor's electrical charges. It is linked to the temperature and electrical resistance of the conductor.
- Temperature noise: This is a variation of the characteristics and parameters of a measuring instrument with temperature due to its instantaneous and long-term variations.
- Shot noise: This is caused by the fact that the carriers of charges are particles and not a continuous flux.
- $1/f$ noise: If an astrophysical signal is detected by a receiver that is a combination of different instruments with different time constants, the detected signal will have slow variations with a spectrum that are proportional to $1/f$.
- Noise of a fundamental origin: This derives from the application of the Heisenberg uncertainty principle. It is impossible to perfectly determine both energy and time with one measurement.
- Photon noise: This is shot noise for photons. These are fluctuations related to photons that arrive at random times. The greater the incident power (either from the target of interest, or causally in front of the receiver), the greater the photon noise. It is often the limiting element in measurements.

8.3.1 Johnson Noise

Johnson noise arises from the fluctuations of electrical charges in a conductor. It is linked to the temperature and electrical resistance of the conductor (Figure 8.5).

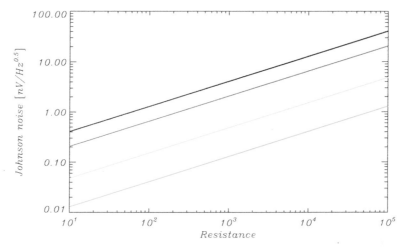

Figure 8.5. Johnson noise as a function of the resistance R and the temperature (colored lines). From top to bottom, the noise at a temperature of 300 K, 77 K, 4.2 K, and 0.3 K.

This is of Brownian origin, and in this case the speed is the current and the mobility is the conductance. We have that

$$\frac{\tau_1}{m} = G. \tag{8.25}$$

The power spectrum of current fluctuations will therefore be white (with all frequencies in it) up to a frequency (bandwidth) of $1/\tau$. The bandwidth is decided by any capacities or inductances in the circuit. The power spectrum will therefore be:

$$S(\omega) = \frac{4k_B T G}{1 + \omega^2 \tau^2}. \tag{8.26}$$

The total power (current) in a limited bandwidth is:

$$\langle \delta I^2 \rangle = \int_0^{\Delta f} S(f) df = 4k_B T G \Delta f \tag{8.27}$$

while the voltage power spectrum is:

$$\langle \delta V^2 \rangle = R^2 \langle \delta I^2 \rangle = 4k_B T R \Delta f. \tag{8.28}$$

A resistance at a temperature T has a voltage noise related to its resistance R and temperature T (Figure 8.5).

In order to reduce Johnson noise, it is necessary to cool down an instrument. Johnson noise is also a fundamental effect that can be used to measure temperatures. We consider a resistance in a circuit on an antenna that is immersed in a blackbody cavity. The emitted power is:

$$P = \sqrt{\langle \delta V^2 \rangle \langle \delta I^2 \rangle} = \sqrt{\frac{R}{R}} 4k_B T \Delta f. \tag{8.29}$$

The power is independent of the resistance and sets a minimum value for a circuit. At 300 K, we have $W = 4 \times 10^{-21}$ W Hz$^{-1/2}$. At 0.1 K we have W $= 1 \times 10^{-24}$ W Hz$^{-1/2}$. These are small values but they are useful for precision calibrations (e.g., for bolometers). Another important point about Johnson noise is the fact that it is based on primary principles, and so it can be used for noise generators (for instance, to measure the Boltzmann constant k_B).

8.3.2 Temperature Noise

Temperature noise arises from both instantaneous and long-term variations of an instrument's characteristics and parameters with temperature. In certain circumstances, this susceptibility is a useful feature, for example in semiconductors or superconductors, whose dependence of characteristics on temperature is exploited. Nevertheless, this dependence has disadvantages. To reduce it, we use feedback systems to create an equal and opposite effect to any temperature change.

These electronic feedback systems must be effective in different ranges of temperature for different applications:
- 0 °C to 70 °C for commercial components
- −25 °C to 80 °C for industrial components
- −55 °C to 125 °C for military components.

Slow effects usually induce $1/f$ noise, while the fast ones will depend on the thermal capacity C of the component considered and on the conductivity with a "bath" at temperature T_0.

The power spectrum is:

$$S(\omega) = \frac{4k_B TG}{G^2 + \omega^2 C}. \tag{8.30}$$

Also, cryogenic techniques are required to minimize temperature noise.

8.3.3 Shot Noise

The shot noise is a consequence of the fact that charge carriers are particles. A typical example of shot noise is the noise (fluctuation) of raindrops on a roof. The effect is noticed when the flow of carriers crosses a potential barrier (e.g., in a diode), but it is not noticed in a conductor. Assuming that each electron is independent of the others, Poisson statistics governs this effect and thus:

$$\langle \delta n^2 \rangle = \langle n \rangle. \tag{8.31}$$

In fact, according to Poisson statistics, the variance and expected value are the same. So, the electrons will have fluctuations as follows:

$$\langle \delta I^2 \rangle = 2e \cdot I \cdot \Delta f. \tag{8.32}$$

Again, this is a flat spectrum that is independent of temperature.

8.3.4 1/f Noise

The astrophysical signal detected by a receiver comprised of different instruments with different time constants will have slow fluctuations with a spectrum that is proportional to $\approx 1/f$ (Figure 8.6). This is linked to a flow of electric charges that, however, is conditioned by multiple effects and instruments (and their lack of ideality). We have seen several noise effects that have a Lorentzian power spectrum characterized by a time constant τ. The time constants in play here, however, are many, so:

$$S_i(\omega) = \frac{A_i}{1 + \omega^2 \tau_i^2} \rightarrow S_{\text{tot}}(\omega) = \sum_i S_i(\omega) = \sum_i \frac{A_i}{1 + \omega^2 \tau_i^2}. \qquad (8.33)$$

Thus, going to a continuous sum, we have:

$$S_{\text{tot}}(\omega) = \sum_i \frac{A_i}{1 + \omega^2 \tau_i^2} \approx \int_0^\infty \frac{A}{1 + \omega^2 \tau^2}. \qquad (8.34)$$

So, if $\tau' = \omega\tau$, we have

$$S_{\text{tot}}(\omega) = \int_0^\infty \frac{A}{1 + \tau'^2} \frac{1}{\omega} d\tau' \propto \frac{1}{f}. \qquad (8.35)$$

The 1/f noise typically originates in carbon resistors (not in metal ones) and in the electrical contacts of metal–oxide–semiconductor field-effect transistors (MOS-FETs). One way to obtain lower levels of 1/f noise for a resistance is to use N resistances in series and in parallel instead of R: the 1/f noise for each resistance will be unrelated and their contribution will sum up quadratically.

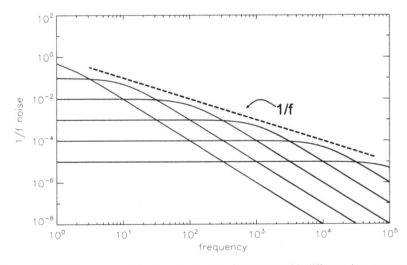

Figure 8.6. 1/f noise as a combination of several noise sources with different time constants.

8.3.5 Fundamental Noise

Noise of fundamental origin arises from the Heisenberg uncertainty principle, which states that it is impossible to determine both position and momentum for a particle:

$$\Delta x \Delta p \geqslant \frac{h}{4\pi}. \tag{8.36}$$

Another formulation concerns energy and time. Fundamental noise is a typical noise for amplifiers that cannot preserve information about both energy and phase (i.e., time):

$$\Delta E \Delta t \geqslant \frac{h}{4\pi}. \tag{8.37}$$

When photons are detected, their energy is $n\nu$ and the wave phase is $\phi = 2\pi\nu t$. Thus:

$$\Delta n \Delta \Phi = \frac{\Delta E}{h\nu} \Delta t 2\pi\nu = \Delta E \Delta t \frac{2\pi}{h} \geqslant \frac{1}{2}. \tag{8.38}$$

Amplification means increasing the energy or the number of photons detected with the consequence of reducing the knowledge about the phase. A "solution" to the problem is to completely lose phase information.

8.3.6 Photon Noise

Photon noise arises from fluctuations related to photons arriving at random times (a shot noise for photons). Photon noise increases with increasing incident power. We try to ensure that it is the limiting factor in measurements. Again, we can apply Poisson statistics:

$$\langle \delta E^2 \rangle = (h\nu)^2 \langle \delta n^2 \rangle = (h\nu)^2 \langle n \rangle = (h\nu)^2 \frac{Wt}{h\nu} = h\nu Wt. \tag{8.39}$$

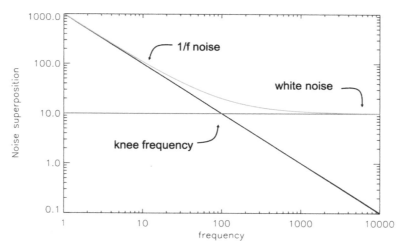

Figure 8.7. Noise in a log ν–log f space.

As mentioned, this is an effect that is proportional to power. Actually, we should use Bose–Einstein statistics; therefore, we will have an additional wave (a bouncing) effect that we will neglect here. In the millimeter band this effect is particularly important because the contributions of the atmosphere, the telescope, and the optics to the power emitted must be considered.

A plot of all of the noise sources we have seen in this chapter is shown in Figure 8.7. We have highlighted the frequency, the knee frequency, below which the $1/f$ noise overtakes the other white noise.

Chapter 9

Cryogenics

The need for cryogenics is highlighted in this chapter. Cryogenic techniques, the thermodynamics of cryogens, wet cryostats versus dry cryostats, mechanical cryocoolers, sub-Kelvin refrigerators, helium-3/helium-4 refrigerators, dilution refrigerators, and adiabatic demagnetization refrigerators are also discussed, as are vacuum techniques. How efficiently one can cool down a certain quantity of copper down to 4K will be calculated as a worked example.

9.1 Introduction

Most of the noise sources we have studied so far depend on the temperature of the device used. Johnson noise, for instance, has lower amplitude at low temperatures. The power spectrum, in tension, at the edges of a resistor of temperature T and resistance R is:

$$S(\omega) = \frac{4k_\mathrm{B}TR}{1 + \omega^2\tau^2}.$$ (9.1)

Also, temperature noise is lower and more stable at low temperatures:

$$S(\omega) = \frac{4k_\mathrm{B}T^2G}{G^2 + \omega^2C^2}.$$ (9.2)

Dark current has also a temperature origin. A CCD camera that observes with the shutter closed still accumulates a dark signal linked to thermal agitation. This must be measured for each pixel (through the use of so-called dark frames) and decreased

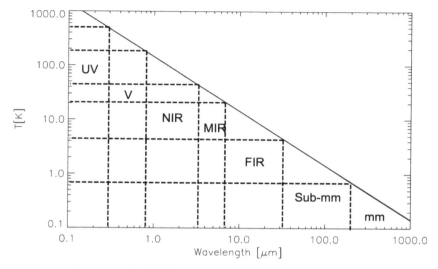

Figure 9.1. Photon noise compared with Johnson noise. At different wavelengths, the approximate required temperature for a detector is indicated. In the UV and optical, we may need to cool detectors down to 77 K. At longer wavelengths, the required temperatures are even lower, down to sub-Kelvin levels for millimeter wavelengths.

by lowering the temperature of the camera to several degrees below 0 °C (modern, very advanced CCDs work at 77 K). Also, in this case there is a clear advantage to lowering the temperature.

Let's take the minimum noise that a detector can have at a given temperature and compare it with the photon noise (one hundredth of it). We have:

$$k_{\mathrm{B}}T \sim \frac{h\nu}{100} \rightarrow T < \frac{h\nu}{100 \cdot k_{\mathrm{B}}} \tag{9.3}$$

which approximately indicates the temperature required for detector noise to be less than the photon noise, as shown in Figure 9.1.

In addition to reducing noise, cryogenic techniques allow you to take advantage of effects due to the behavior of matter that are otherwise masked by thermal agitation. These are, for instance, superconductivity and superfluidity, which become prominent when a material approaches the critical temperature T_{c} (Figure 9.2).

In addition to the behavior of magnetism at cryogenic temperatures, electrical, thermal, mechanical, and optical properties of materials all vary with temperature. All of these properties can be an advantage and a disadvantage at the same time. Cryogenics is a complete science because it combines the knowledge of many different fields.

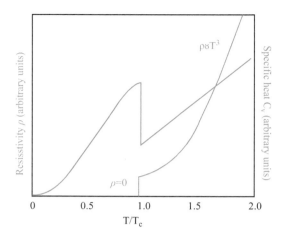

Figure 9.2. Heat capacity (C_v, red) and resistivity (ρ, blue) at the superconducting transition.

Table 9.1. Liquid Cryogens Used in Cryogenics with Their Boiling Temperatures

Cryogen	Boiling Point [K]
Methane	111.7
Oxygen	90.18
Argon	87.24
Fluoride	85.24
Air	78.8
Nitrogen	77.36
Neon	27.09
Hydrogen	20.27
Helium-4	4.217
Helium-3	3.19

9.2 Cryogens (Liquids)

The easiest way to cool down a detector and keep it at a constant temperature is to put it in contact with a boiling liquid (or a sublimation solid) that has a low boiling temperature (Table 9.1).

The latent heat of evaporation allows the temperature of a detector to be kept constant (at the expense of the consumption of the cryogen). Commonly used cryogens are listed in the following, and the most used are nitrogen and helium-4, whose latent heat and boiling point are listed in Table 9.2:

- Oxygen
- Nitrogen

Table 9.2. Nitrogen and Helium Boiling Point and Latent Heat

Cryogen	Boiling Point [K]	Latent Heat [kJ l^{-1}]
Nitrogen	77.37	161
Helium	4.217	2.6

- Neon (very expensive)
- Hydrogen (explosive)
- Helium.

Once the system is thermalized at 77 K and then at 4.2 K, if we want to drop the temperature even further, we can reduce the saturated vapor pressure above the cryogenic bath. This is done, in the laboratory, using vacuum pumps that pump on the cryogenic bath. However, this process consumes a considerable additional amount of liquid (part of the liquid evaporates to cool the rest of the liquid). This can be calculated as follows. Given the specific heat C (T) of the cryogen, which depends on the temperature, and its latent heat $L(T)$, we have:

$$L(T)dm = mC(T)dT \rightarrow \log\left(\frac{m}{m_0}\right) = \int_{T_0}^{T} \frac{C(T)}{L(T)}dT. \tag{9.4}$$

The pressure on the bath can be derived from the Clausius–Clapeyron law:

$$\frac{dP}{dT} = \frac{L}{T(V_V - V_L)}. \tag{9.5}$$

The expansion ratio (the ratio between the volume of vapor and that of the liquid) of cryogenic liquids is usually 700 (696 for nitrogen, 757 for helium), and thus:

$$\frac{dP}{dT} = \frac{L}{TV_V} \cong \frac{PL}{RT^2} \rightarrow P \propto e^{-\frac{L}{RT}} \tag{9.6}$$

which means that, even after lowering the pressure on a bath, each liquid has a limit dictated by the exponential law just shown.

Worked Example: How to efficiently cool down 100 kg of copper.
  Suppose we want to cool down 100 kg of copper from 300 K to 4.2 K. How much do money will it cost?
  For copper we have the following heat capacity at high temperature (HT, above 77 K) and at low temperature (LT, below 77 K):
  - $C^{HT}_{Cu} = 300\ J\ (kg\ K)^{-1}$
  - $C^{LT}_{Cu} = 100\ J\ (kg\ K)^{-1}$

The latent heat for Helium and Nitrogen are:
  - $L_{He-4} = 2.6\ kJ\ l^{-1}$
  - $L_{N2} = 161\ kJ\ l^{-1}$

And the price is approximately:
  - $^4He = 9.5€\ l^{-1}$
  - $N_2 = 0.8€\ l^{-1}$

We can use only liquid helium-4. In this case we need the following heat:

$$Q = C_{Cu}m\Delta T \simeq 9\ MJ \rightarrow \frac{Q}{L\ ^4He} \cong 3400\ l$$

which yields, with the price per liter of helium, a cost of 33,000€.
  However, we can pre-cool with liquid nitrogen and then cool with helium-4. In this case the heat is:

$$Q_{300\ K-77\ K} = C^{HT}_{Cu}m\Delta T \simeq 6.7\ MJ \rightarrow \frac{Q_{300\ K-77\ K}}{L_{N_2}} \cong 41\ l$$

$$Q_{77\ K-4.2\ K} = C^{LT}_{Cu}m\Delta T \simeq 0.7\ MJ \rightarrow \frac{Q_{77\ K-4.2\ K}}{L\ ^4He} \cong 280\ l.$$

So, given the cost of liquid helium and nitrogen, the total price is

$$33€ + 2660€ \approx 2700€.$$

This is clearly cheaper than the previous case. It is thus important to pre-cool. One should mention, however, that in reality we should also consider enthalpy (which for helium is greater than the latent heat) and perform the integral over specific heat. Pre-cooling is necessary to save money and time!

9.3 Thermal Inputs

Once an experimental apparatus has been cooled to a cryogenic temperature, the temperature is maintained at the expense of the consumption of the liquid cryogen. Consumption is caused by the thermal input. Due to the low latent heat, it is not possible to keep a cryogen in a normal container, and we need a container with low thermal inputs. The thermal input $\dot{Q}$ is what dictates the duration of the cryogen.

This of course depends also on the latent heat L [J l^{-1}] and on the volume V [l] of the liquid. The time t that a cryogen will last will thus be:

$$t = \frac{LV}{\dot{Q}}. \qquad (9.7)$$

The thermal input can have different origins:
- Conduction: linked to the transport of heat by contacting surfaces
- Irradiation: linked to the exchange of radiative heat between bodies at different temperatures
- Convection: linked to the transport of heat by means of a gas that keeps two stages with very different temperatures in thermal contact.

In addition to the thermal input of the radiation of astrophysical interest. A container that keeps detectors cold is called a cryostat (a Dewar vessel) and is a system of concentric cylindrical containers designed in such a way as to contain the detectors (and place them in the focal plane of a telescope) by means of suitable supports and while minimizing all thermal inputs.

9.3.1 Conduction

The conductive input is due to the mechanical supports of the individual Dewar parts and detectors, as well as to the electrical cables used for reading the signals. The "tanks" of cryogenic liquids are typically attached to the "neck" of the tank itself (the tube through which liquids are transferred). The thermal input due to conduction depends on the geometric and thermodynamic properties of the materials used. Given a bar of length L, surface S, and thermal conductivity k, we have

$$\dot{Q} = k\frac{S}{L}(T_2 - T_1). \qquad (9.8)$$

Since usually the thermal conductivity depends on the temperature, $k = k(T)$, we have:

$$\dot{Q} = \frac{S}{L}\int_{T_1}^{T_2} k(T)dT. \qquad (9.9)$$

In practice it is sometimes useful to consider an average $<k(T)>$ defined as:

$$\langle k \rangle_{T_1 - T_2} = \int_{T_1}^{T_2} k(T)dT. \qquad (9.10)$$

Therefore, it is necessary to minimize S (and k) and to maximize L. Furthermore, it is advisable to create various intermediate stages at gradually decreasing temperatures in order to "load" the inputs gradually. When choosing the geometry and materials to use, a trade-off between mechanical strength and conductivity must be considered.

Copper is the metal with the highest conductivity, so it is appropriate for use in thermal links, but not in supporting structures. There are several varieties, including electrolytic copper and oxygen-free high conductivity (OFHC) copper. Steel and aluminum are much better materials for support structures. Alternatively, plastic materials such as Vespel, Teflon, Vetronite, Kevlar, Torlon, Invar, and G-10 have very low thermal conductivity values and are also strong.

9.3.2 Irradiation

Irradiation is caused by radiative heat exchange between bodies at different temperatures. The thermal input by radiation from a hot surface to a cold surface can be obtained by remembering that the total power emitted by a body at temperature T, of area A, and emissivity ε, on 4π solid angle is:

$$\dot{Q} = \varepsilon\sigma A T^4. \tag{9.11}$$

So, if we consider two surfaces facing one another at temperatures T_1 and T_2 we have:

$$\dot{Q} = F_{12}\sigma A_1(T_2^4 - T_1^4) \tag{9.12}$$

where the empirical term F_{12} is a factor that depends on the geometry (weakly, on the area) and indicates how much of the surfaces are "facing" each other. For concentric surfaces very close to each other and of equal area (a good approximation for a cryostat), we have:

$$F_{12} = \frac{\varepsilon_1\varepsilon_2}{\varepsilon_1 + \varepsilon_2 - \varepsilon_1\varepsilon_2} \xrightarrow{\varepsilon_1=\varepsilon_2=\varepsilon\ll1} F_{12} \approx \frac{\varepsilon}{2}. \tag{9.13}$$

In order to reduce the irradiation input, the dimensions must therefore be minimized, but they are often dictated by the experiment (the optics, angular resolution, the number of detectors, etc.). The best way to reduce the radiative thermal input is therefore to decrease the emissivity by using suitable materials or polishing the surface of a shield. Also, in this case, it is useful to create intermediate stages both to reduce the effect on the lower temperature stage and to prevent the entrance from heating the surfaces. In addition, we can create a so-called "super-insulation" or a powder insulation that drastically reduces the thermal input:

- Superinsulation: A high number n of intermediate screens, reduces the irradiated power. Usually, this is done with many sheets of aluminated Mylar, which is insulating, has low ε, and is slightly corrugated to avoid firm contact. This is often called multilayer insulation (Figure 9.3).
- Powder insulation: The insertion of small grains in the cavity allows the radiation to spread from one screen to another and therefore increases the path of the radiation (and decreases F_{12}). However, this method is efficient if the vacuum does not have to be very low.

9.3.3 Convection

Convection thermal input arises from the transport of heat by a gas that keeps two materials with different temperatures in thermal contact with each other. The thermal input by convection is similar to conductive input, but it is linked

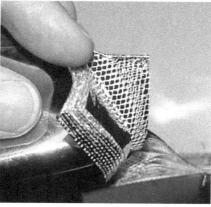

Figure 9.3. Multilayer insulation (MLI). Left: internal shield coated with MLI. Credit: Wikipedia: TorFraTe (CC BY-SA 4.0). Right: a sample of superinsulation. Credit: Wikipedia: Dantor assumed (CC BY-SA 2.5).

to the pressure p of the gas present in the interspace. It can be quantified as follows:

$$\dot{Q} = Ka_0 p(T_2 - T_1) \qquad (9.14)$$

where the coefficient K depends on the gas ($K = 2.1$, 4.4, and 1.2 for helium, hydrogen, and air, respectively), and a_0 depends on the geometry (similar to F_{12}). To reduce these effects, cryostats are built using the same principle as for a thermos. There are two concentric containers: the inner one is the cold one (with the liquid and detectors), and the outer one is at room temperature. In the interspace between the two containers, a vacuum is made. Clearly, a cryostat must be kept vacuum tight and needs to be checked for leaks.

In the mechanical design of a Dewar, it must be taken into account that the external pressure is very high, e.g., 10,000 kg m^{-2}. The materials used for the external structure (shell) of a cryostat are steel or reinforced aluminum. Given the need to mount the instrument inside it, it is necessary to provide flanges that hold the vacuum. Typically, rubber O-rings are used to seal the flanges, and their slots must be designed while keeping in mind the working temperature (see the 1986 Space Shuttle disaster[1]). Among the materials used, we have (Figure 9.4):

- Butyl, down to −60 °C (but it is permeable to helium)
- Viton, down to −20 °C
- Metal O-rings (gaskets), which are used in an ultra-high vacuum
- Ductile metal O-rings (e.g., indium), which are used in cryogenics.

---

[1] https://www.nasa.gov/multimedia/imagegallery/image_gallery_2437.html

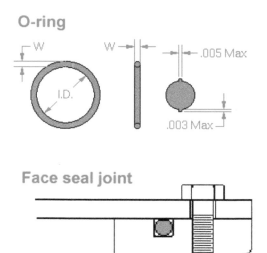

Figure 9.4. O-ring (top) and its cave of a cryostat (bottom). Credit: Wikipedia (Public Domain) https://commons.wikimedia.org/wiki/File:O_ring.png.

9.4 Vacuum Pumps

In order to reduce the effects of convection from the entrance, the cavity of a cryostat must be evacuated. Vacuum techniques have countless applications in research and in industry. Vacuum levels are divided into three categories:

- Low vacuum: 10^3 mbar $> P > 10^{-2}$ mbar. At this pressure, a container contains more molecules than are deposited on its surfaces.
- High vacuum: 10^{-2} mbar $> P > 10^{-6}$ mbar. Here, most of the molecules are localized on the container surfaces.
- Ultra-high vacuum: 10^{-6} mbar $> P$. Here, the surfaces are, in all respects, clean.

To evacuate the cavity of a cryostat, the internal walls first must be cleaned (typically with volatile products with a high vapor pressure, such as alcohol), avoiding the use of water. Then the container to be evacuated is connected to a vacuum pump through a connection line (which is characterized by an impedance linked to the diameter of the flanges and bellows). In the case of a high or ultra-high vacuum, the efficiency with which it is evacuated is therefore very important, and the diameter of the interconnection is important. In the case of a low vacuum, during the evacuation the gas molecules meet other molecules, and therefore the impedance of the emptying line is less important (Figure 9.5).

In any emptying regime, the gas flow (V versus t) in a tube is described in terms of pumping speed Q, which depends on the pressure p:

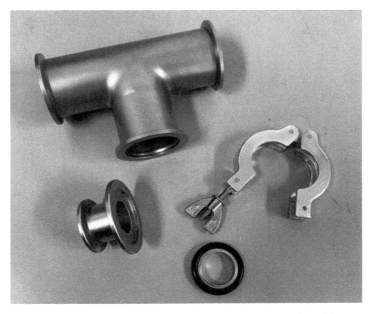

Figure 9.5. Vacuum flanges used to make a vacuum. A KF-25 tee and an O-ring with a centering ring and clamp are shown. Credit: Wikipedia: Kkmurray (CC BY-SA 3.0).

$$Q = p\frac{dV}{dt}. \tag{9.15}$$

We can define the total conductance of a pumping line that has p_1 and p_2 pressures upstream and downstream of the line as the following:

$$C = \frac{Q}{p_1 - p_2}. \tag{9.16}$$

If we add another line to a first one in parallel with it, we have that the total conductance is:

$$C = C_1 + C_2 \tag{9.17}$$

while if we connect two lines in series, we have that:

$$\frac{1}{C} = \frac{1}{C_1} + \frac{1}{C_2}. \tag{9.18}$$

Conductance C is measured in $l\,s^{-1}$ and is an indication of how efficient a connection line is in evacuating a container. In the different regimes, we have different dependencies on the line diameter D and on the temperature T.

The vacuum pumps are divided into low-vacuum pumps and high- or ultra-high-vacuum pumps. The former is used to bring the latter to operating conditions of 10^{-1} mbar. They can also be divided by cleaning and the continuity of operations:

- Low-vacuum pumps, 10^3 mbar $> P > 10^{-3}$ mbar:
  - Rotary pumps
  - Booster and scroll pumps
  - Sorption pumps (cryogenic)
- High-vacuum pumps, 10^{-1} mbar $> P > 10^{-6}$ mbar:
  - Oil diffusion pumps
  - Turbomolecular pumps
- Ultra-high-vacuum pumps, 10^{-5} mbar $> P$:
  - Sublimation pumps
  - Ionic pumps
  - Getter pumps.

9.4.1 Low-vacuum Pumps: Roughing Pumps

One of the most popular roughing pumps is the rotary pump. Its operating principle is the following: an eccentric cylinder is rotated by means of a motor. The rotation compresses and decompresses the gas. The vacuum seal is dynamic and is covered by a layer of oil. The gas to be evacuated is made to enter through an entrance door and is compressed by the motion and directed toward an exit door (Figure 9.6).

Limiting pressures for rotary pumps are on the order of 10^{-2}–10^{-3} mbar. At the exit they have exhausted oils. They must have check valves that prevent the oil from returning to the evacuation chamber. Capacities are of order 1–50 l s^{-1} for small air-cooled pumps and up to 500 l s^{-1} for water-cooled pumps.

Less powerful than a rotary pump but with advantages related to the non-use of oils are the booster and scroll pumps (Figures 9.7 and 9.8). Boosters have two rotors that are rotated synchronously, leaving a gap of 0.1 mm. They fail to drain over large pressure ranges and are sometimes used in a cascade configuration. They are typically used to mix helium-4 and helium-3 in a dilution refrigerator (see Section 9.7.2).

Scroll pumps have springs that compress the gas thanks to the rotation of one of them. They are used as a backing pump (Figure 9.8).

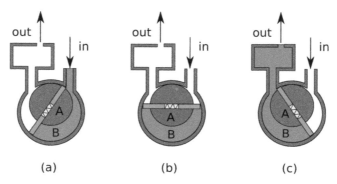

Figure 9.6. Rotary pump: an eccentric rotor (A) rotates in a stator (B) and allows the space to be evacuated in three phases: (a) the fluid is input into the pump, (b) it is transported toward the output, and (c) it is output from the pump. Credit: Wikipedia: Di Daniele Pugliesi—Pompa rotativa.png (Antonio84; CC BY-SA 3.0).

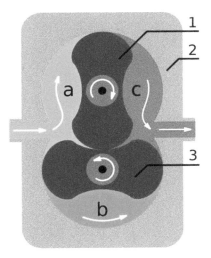

Figure 9.7. Booster pump: the fluid gets trapped in pockets surrounding the lobes and carried from the input to the output by the lobes' rotation. Credit: Wikipedia: Inductiveload (CC BY-SA 3.0).

Figure 9.8. Scroll pump. The rotation of the blades compresses the fluid and outputs it. Credit: Wikipedia: Von Albert Handtmann Metallgusswerk GmbH & Co. KG / Handtmann01 in der Wikipedia auf Deutsch—Eigenes Werk (CC BY-SA 3.0).

In sorption pumps, the gas is trapped by a very porous material (activated carbon whose surface-to-volume ratio is very high: $S/V \sim 700$ m^2 cm^{-3}; Figure 9.9), which adsorbs it due to the molecular cohesion forces that lose kinetic energy when it is cooled and releases the gas when it is heated. They work only in a small range of pressures and must be cooled depending on the gas to be removed. An advantage is that they are intrinsically passive cryogenic pumps, but they must be cooled and heated. A disadvantage is that they cannot be used continuously, so they are sometimes used in successive stages (after the first is saturated, the second is activated while the first is regenerated).

9.4.2 High-vacuum Pumps

A diffusion pump, which is used only after pre-evacuating a chamber, can operate roughly, from 10^{-2} mbar and reach pressures of 10^{-7} mbar. A jet of oil vapor drags

Figure 9.9. Activated carbon. This is the material used to build sorption pumps. Credit: Wikipedia: Self (CC BY 2.5).

the gas molecules to be evacuated toward the exit. The oil is heated, and its vapors are directed with high speed toward the outlet. Then the diffusion pump must be connected to an auxiliary pump (a backing pump) that extracts the molecules from the system. All of these pumps have a big backscatter problem, so cryogenic traps are used at the inlet to the diffusion pump (Figure 9.10).

A turbomolecular pump is a "clean" high-vacuum pump with a purely mechanical operating mechanism. Limiting pressures are $\sim 10^{-9}$ mbar starting at 1 mbar, and these pumps also must be followed by backing pumps that removes the extracted gas. A turbine gives the molecules to be evacuated a momentum toward the exit. Turbomolecular rotors also rotate at 50,000 rpm and some are magnetic levitation turbines (Figure 9.11).

9.5 Cryostats and Their Cool Down

A cryostat is a Dewar in which we place our detectors to cool them down. The process of cooling down a liquid cryogen Dewar involves a certain number of operations and steps to ensure the cool down is successful. They are listed in the following:

1. Mount the instrument (typically, it is well anchored to the cold flange).
2. Close the cryostat:
   a. Clean the internal surfaces.
   b. Grease the O-rings with silicone greases.
   c. Tightly close all flanges by closing the screws in a "star" configuration.
3. Attach the vacuum lines to the vacuum valve.
4. Start pumping with the roughing pump (e.g., a rotary pump).
5. Once a high-vacuum level is reached, change to a high-vacuum pump (e.g., a diffusion or turbomolecular pump).

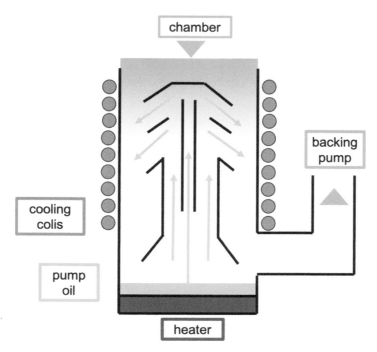

Figure 9.10. Diffusion pump.

Figure 9.11. Turbomolecular pump. Credit: Wikipedia: Liquidat (CC BY-SA 3.0).

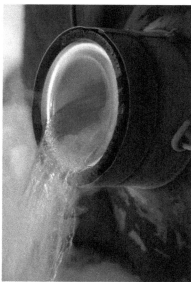

Figure 9.12. Left: a cryostat used to store liquid nitrogen. Credit: Wikipedia: Di Jeffrey M. Vinocur—Opera propria (CC BY 2.5). Right: liquid nitrogen being poured. Credit: Wikipedia: Par Robin Müller (CC BY-SA 3.0).

6. Before transferring liquids, check that there are no leaks by attaching a "leak detector" to the vacuum valve. A leak detector is a pump (usually a roughing pump + high-vacuum pump in series) followed by a mass spectrometer that detects helium. In a mass spectrometer, the molecules are ionized, accelerated, and then, through orthogonal magnetic and electric fields, they are selected based first on their speed, and then on their mass/charge ratio. The ratio to that of the helium-4 is selected and the gaskets, flanges, and all of the critical parts of the Dewar are sprayed with helium-4 gas, from top to bottom, checking first that the sensor of the mass spectrometer does not detect anything. The leaks are measured in mbar l s^{-1} < 10–11.

7. Transfer the nitrogen by pouring it (carefully) or using transfer lines inserted directly into the descendants of the Dewar. Its high latent heat of evaporation implies that a container (that is perhaps made of polystyrene or has a double wall) can contain nitrogen for a long time. All tanks in a Dewar must be cooled to 77 K. Nitrogen is not particularly dangerous, but we must remember:
   i. Not to touch it
   ii. Not to close it in a sealed container (expansion ratio ~700)
   iii. Not to saturate a room with it

During the transfer, the immediate evaporation that occurs at room temperature means that the liquid is not "accepted" in a short time due to oscillations (Figure 9.12).

8. Once the cryostat has thermalized to 77 K (which sometimes takes a long time), perform another leak test. This test is more effective than the first (it reveals any smaller leaks) because, with nitrogen, the walls of the cryostat have adsorbed any molecules left in the cavity.

9. At this point, it is necessary to remove all of the nitrogen in the internal tank (this is important because the latent heat of nitrogen fusion is higher than that of helium and therefore all of the helium would evaporate).

10. Use a tube that reaches the bottom of the tank to pressurize the bath with helium-4 gas. If the nitrogen has been completely removed, the temperature should rise above 77 K. Liquid helium is more difficult to transport: transport Dewars equipped with superinsulation, pressure gauges, and safety valves (and wheels) must be used. A transport Dewar has valves to insert helium-4 gas. Always remember, never close the relief valves.

11. Use a transfer line to transfer the helium from the transport Dewar to the cryostat. Due to the low latent evaporation heat of helium, a transfer line has a double wall with a void between the walls (otherwise the contents would evaporate and, freeze). The transfer line (which can be rigid or flexible) is sealed in the transport Dewar, which is pressurized. In order to "convince" the liquid helium to go from one container to another, its surface can be pressurized by:
    i. Simply the insertion of the transfer line at the beginning of the transfer
    ii. The insertion of helium gas
    iii. A balloon and some mechanical energy.

12. During the first transfer (immediately after the nitrogen is removed), insert the helium transfer line down to the bottom of the tank in order to exploit the enthalpy of helium (so that the fumes of evaporation pre-cool the tank). For second and subsequent transfers of "re-fill" helium, the opposite is valid: this procedure prevents "smoking" by any liquid already deposited. The same principle is used in cryostats that are equipped with intermediate screens and cooling tubes on them.

In general, each cryostat has many small "tricks" that require a thorough knowledge of its behavior. Problems can arise when an astronomer fills it in a difficult environment, such as a telescope dome.

9.6 Mechanical Cryocoolers

Mechanical cryogenerators are an alternative to cryostats with cryogenic liquids. Their advantages over liquid refrigeration are ease of use (especially in remote areas, like Antarctica) and relative safety. The disadvantages are their electricity consumption (>10 kW; this precludes their use on space platforms) and mechanical vibrations. In all cases, cryocoolers work through (hot) compression and (cold) decompression of a gas, super-pure helium, that extracts heat from a system (Figure 9.13).

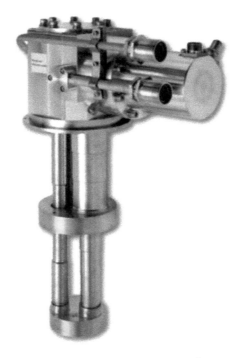

Figure 9.13. The cold head of a mechanical cryocooler. This Sumitomo[2] model cools down to 4 K with a two-staged cryocooler: a 40 K stage and a 4 K stage.

There are several types of cryocooler, including Stirling, Gifford–McMahon, and pulse tube ones. They differ according to the kinds of cycles used and the moving parts of the cold system. In general, a cryocooler consists of at least two units with strong seals designed to work at high pressures that are connected via pipes: an external unit to the cryostat: a compressor that compresses the gas; and a cold head with a valve, which is where the cold is produced.

The cold finger is a unit inside the cryostat, connected to the "head" of the cryocooler where the gas expands. In some cases, there is an exchanger to cool the compressor through water or air cooling.

In a Stirling cryocooler, there are: two pistons, one at temperature T_a and one at T_L; two heat exchangers, one at T_a and one at T_L, which transfer and store heat at constant T; and a regenerator, which is a structure through which the gas passes, maintaining a constant volume, and which is formed of a porous material with a great thermal capacity (a high thermal conductivity with the gas but a low thermal conductivity in the direction of the gas flow; Figure 9.14). The regenerator absorbs and transfers heat to the gas that passes through it. It is assumed that the thermal contacts with the zones at $T = T_a$ and $T = T_L$ are perfect; therefore, the compressions and expansions of the pistons are isothermal. The passage through the regenerator is instead an isochor-like transformation.

---

[2] http://www.shicryogenics.com/

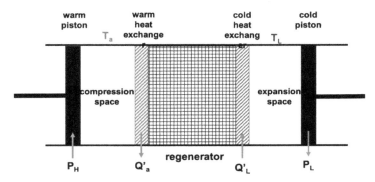

Figure 9.14. Scheme of a Stirling cryocooler.

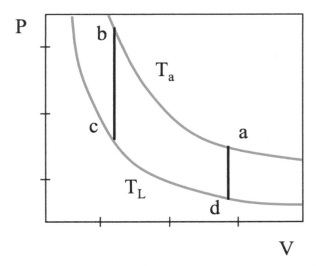

Figure 9.15. Typical thermodynamic cycle performed by a Stirling cryocooler in a Clapeyron plane.

Cooling is described through four phases in the P–V plane (Figure 9.15). The cycle takes place counterclockwise (as for a heat pump; Figure 9.16):

- a–b: The P_H piston performs a hot compression. This compression takes place in an isothermal way, and therefore the heat Q_a escapes from the system.
- b–c: Both pistons move by transferring the gas from the hot zone to the cold zone in an isochor-like way through the regenerator.
- c–d: The P_L piston expands isothermal gas, then the heat enters.
- d–a: Both pistons move in synchronous mode and gas (which has expanded) crosses the regenerator.

Having a piston in the cold part is actually not practical for issues related to mechanical and thermal contractions. A displacer is then used that, in practice, moves the gas by moving the regenerator between T_a and T_L. In a Gifford–

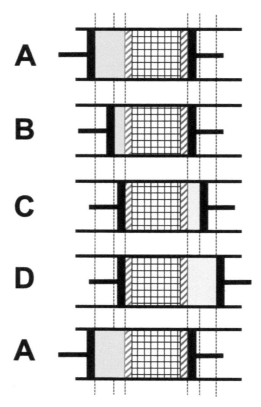

Figure 9.16. The four phases of the thermodynamic cycle of a Stirling cryocooler.

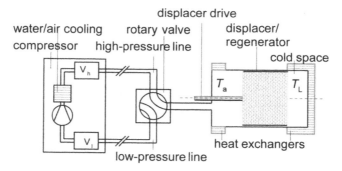

Figure 9.17. Gifford–McMahon cryocooler. Credit: Wikipedia: Adwaele—made by SliteWrite (CC BY-SA 3.0).

McMahon cryocooler, the cold head has both a compression and an expansion zone, and the connection with the compressor takes place via a rotary valve that alternately connects the cold head with a high-pressure and a low-pressure line (Figure 9.17).

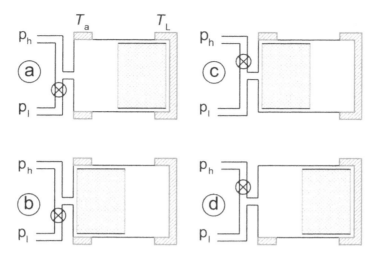

Figure 9.18. Gifford–McMahon cycle. Credit: Wikipedia: Adwaele—made by SliteWrite (CC BY-SA 3.0).

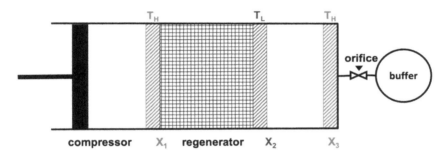

Figure 9.19. Schematic of a pulse tube cryocooler.

The cycle can again be explained in four phases (Figure 9.18):
- a–b: The rotary valve is oriented toward the high-pressure line. The displacer moves the regenerator to the left and the gas at high pressure passes through it.
- b–c: The valve rotates toward the low-pressure line and the gas flows through the regenerator and expands (now cold) isothermally.
- c–d: The regenerator moves to the right and the gas passes through it.
- d–a: The valve rotates toward the high-pressure line and hot compression occurs.

In a pulse tube cryocooler there is a compressor, a regenerator, and also a line with an adjustable impedance orifice (Figure 9.19). It has no cold moving parts. The compressor performs compressions and rarefactions of the gas through a piston. The system is isolated, so the compressions and expansions are adiabatic. The stages of a pulse tube cryocooler are four:
- With the orifice closed, the piston compresses the hot (T_H) gas.

- Subsequently, the orifice opens and the gas flows through the regenerator toward the buffer, and at the same time the piston continues to move to the right (more slowly) to keep the pressure constant.
- The orifice closes and the piston moves to the left; the pressure and temperature drop.
- As before, the orifice opens and the piston moves slowly to the right in order to maintain the constant pressure.

A Joule–Thomson (J–T) cooler has no gas oscillations but a continuous flow: the gas enters at a temperature of 300 K and a pressure of 1 bar and flows at 300 K and 200 bar through the exchanger and is pre-cooled. An expansion occurs through a J–T valve, then the gas returns to a pressure of 1 bar but with a much lower temperature. Part of the gas can leave if it has liquefied. The remainder passes through the exchanger and restarts the cycle. J–T coolers are used as liquefiers and, in specific applications, can be miniaturized.

9.7 Sub-Kelvin Refrigerators

Sub-Kelvin refrigerators allow a detector to be cooled down to well below 1 K, typically to 0.3 K or 0.1 K, depending on the technique used. Temperatures below 1 K are required for detectors working in the far-infrared, the submillimeter, millimeter, and microwave bands. A temperature of 1 K is the lowest that can be obtained by reducing the saturated vapor pressure on a helium-4 bath. To further reduce the temperature, one can use helium-3 (with $T_{ev} = 3.2$ K at 1 atm), an isotope of helium-4 that, however, has a latent evaporation heat of 1 kJ l^{-1} and is more expensive than helium-4 (by a factor of ~10,000).

9.7.1 Helium-3 Refrigerators

The tolerable thermal inputs on a helium-4 refrigerator are on the order of several mW, while for helium-3 this cannot exceed tens of μW (Figure 9.20). The pumping on the helium-3 bath is carried out through cryopumps with cooled activated carbon below 5 K, which allows for pressures of 10^{-3} mbar to be reached, along with a temperature of ~250 mK, depending on the thermal input on the bath. The

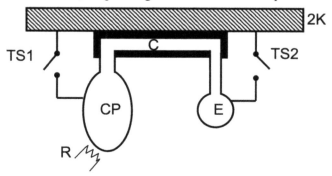

Figure 9.20. Helium-3 refrigerator.

temperature is maintained as long as the helium-3 has not been completely pumped away or the cryopump is not saturated. A helium-3 refrigerator can be completely sealed (with the enormous advantage of not dispersing the helium and being able to reuse it for subsequent cryogenic cycles), and it is composed of two thermal switches (TS), a cryopump (CP), a condenser (C), and an evaporator (E). Suppose we cooled to 2 K with pumped liquid helium-4. The cycle is outlined in the following. The closed circuit contains helium-3 at a pressure of about 100 bar.

- The TS1 switch is open, TS2 is closed, and the CP heats up to about 30 K, allowing the helium-3 to desorb. Helium-3 condenses in C.
- The helium-3 "falls" into the evaporator and when it is all condensed, the CP stops heating, TS2 opens, and the CP starts to cool by closing TS1.
- Once the CP is cold it begins to pump on the helium-3 bath, reaching 250 mK, depending on the thermal input on the bath.

9.7.2 Dilution Refrigerators

A dilution cooler allows temperatures below 300 mK (and even down to 2.5 mK) to be reached. Its working principle makes use of a mix (a dilution) of helium-3 and helium-4. Below a certain temperature, helium-4 becomes a superfluid (as does helium-3, but too low a temperature). The characteristics of a superfluid are the following:

- No viscosity
- Infinite thermal conductivity
- Motion without loss of kinetic energy (perpetual)
- Discontinuity in the specific heat curve shaped in a peculiar way (a "lambda" shape).

Superfluidity is related to the state of a Bose–Einstein condensate. Helium-4 is made up of bosons, and at 2.17 K it becomes a superfluid. Helium-3 is formed of fermions, and to become superfluid, it must form fermion pairs. This happens at a much lower temperature, ~2.5 mK, and is not of interest here.

Suppose we mix helium-4 with helium-3 (e.g., 50% and 50%). Above 2.17 K this mixture will be a "normal" fluid in which helium-3 "dissolves" in helium-4. Below 2.17 K the helium-4 can become a superfluid and, depending on the temperature, the mixture may be composed of helium-3 in superfluid helium-4. With a further decrease in temperature, the mixture separates into two components, one rich in helium-3 (in a normal fluid), and one poor in helium-3 (in a superfluid). At $T < 50$ mK, observing the phase space on the left side of Figure 9.21, it can be seen that the mixture is practically all superfluid helium-4 and is only about 6.6% helium-3. On the right-hand side, it is seen that helium-3 is practically pure. Since helium-3 is lighter, the helium-3-rich phase will position itself on the surface relative to the helium-4-rich phase. Normally helium-3 would "evaporate" from the rich phase to the low helium-3 phase, and the temperature would rise. However, if helium-3 is removed (pumped away from the bottom of the mixing chamber) from the diluted phase (left), the temperature is lowered, as when pumping on a bath (Figure 9.21).

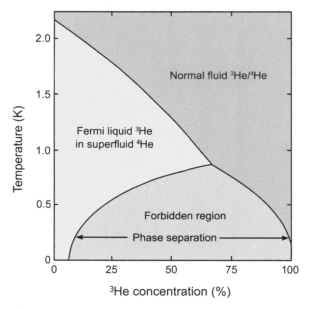

Figure 9.21. Mixture of helium-3 and helium-4 at different temperature as a function of the helium-3 concentration. Credit: Wikipedia: Di Mets501—Opera propria (CC BY-SA 3.0).

In a continuous cycle configuration, we have instead that helium-3 is pumped from the diluted phase and brought to room temperature. It is then reinserted into the cryostat by passing through a nitrogen trap and a helium-4 trap. It is further cooled in contact with a pumped helium-4 bath and is liquefied. Once cooled further, it enters the mixing chamber. Helium-3 evaporates from the rich phase to the poor phase.

9.7.3 Adiabatic Demagnetization Refrigerators

An adiabatic demagnetization refrigerator works by lowering the entropy of a system (for example of paramagnetic salt) with magnetic moments through a magnetic field (~10 T) and then gradually lowering the magnetic field. It is similar to a sorption refrigeration cycle, but instead of increasing and decreasing the pressure, it increases and decreases the magnetic field. Magnetic fields tend to align paramagnetic materials, and thermal motion tends to de-align them. By applying a strong magnetic field to a magnet held, for example, at 4.2 K, the system is ordered, entropy is reduced, and if the salt is in contact with a thermal bath, the temperature remains constant (at the expense of evaporation of the cryogen, e.g., helium-4, which would avoid the natural tendency to raise the temperature). Then, the magnet is disconnected from the bath of helium-4 using a heat switch and the magnetic field is removed in an adiabatic way (a thermally isolated way). If this last step were not adiabatic, entropy would increase, and the grains of salt would misalign. Instead, in this adiabatic transformation, entropy remains constant and temperature decreases.

Experimental Astrophysics

Elia Stefano Battistelli

Chapter 10

Spectroscopy

The importance of spectroscopy will be introduced in this chapter. Low-resolution spectroscopy, high-resolution spectroscopy, and broadband photometry will be described. The working principles of filters and spectrographs, such as the prism and gratings, will be given, along with a worked example related to anti-reflection coatings. Gratings and their use, grisms, and Fourier transform spectrometers will be explained.

10.1 The Importance of Spectroscopy

Most astrophysical information is obtained from spectral measurements. It is in fact important to derive the specific flux or brightness. Among the information that an astronomer can gather through spectroscopic observations, we have: the chemical composition of a source and its state and temperature, redshift, distance, peculiar motion, and kinematics. Astrophysical sources can be characterized through their spectral energy distributions (SEDs), which are often the only way to separate the different contributions to and understand the physical origins of the detected emission by comparison to models. Solar spectroscopy made it possible to discover, in the 19th century, that the Sun was made mostly of hydrogen. In 1868, a new element was discovered from the solar spectrum: helium. Spectro-photometry can be done both in emission and absorption. One can make use of light from a quasi-stellar object (QSO) in the background to measure how the abundances of elements change as a function of redshift. Another important piece of spectroscopic information comes from observations of the 21 cm line, the atomic transition of a hydrogen atom between two states, one with the spins of the electron and of the nucleus parallel and one with the two spins anti-parallel.

doi:10.1088/2514-3433/ac0ce4ch10

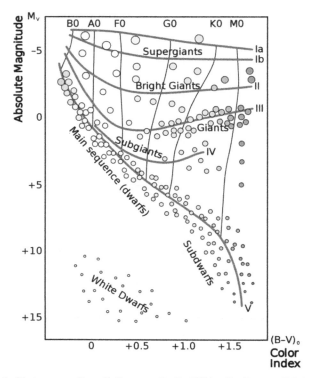

Figure 10.1. Hertzsprung–Russell diagram. Credit: Wikipedia: Rursus (CC BY-SA 3.0).

Today, the Harvard spectral classification of stars is standard. This classification scheme is based on a relation between color and magnitude and was conceived by Father Angelo Secchi in 1866 (Figure 10.1).

Stars have temperatures between 3000 K and 40,000 K and their temperatures are related to their color index (B–V magnitude). The blackbody emission of a star is transmitted by the "cold" stellar atmosphere (the photosphere), which leaves its traces in absorption. The radiative transport equation allows one to study its physics. The stars can be divided into different classes (temperature are approximate):

- O: blue stars, $T = 25{,}000/30{,}000$–$40{,}000$ K, B–$V < -0.2$, HeII ionization lines, HeI lines are visible, weak HI lines. Examples: Zeta Puppis, 10 Lacertae.
- B: blue–white stars, $T = 10{,}000/15{,}000$–$25{,}000/30{,}000$ K, $-0.2 < B$–$V < 0.0$, the HeII lines disappear, the HeI lines are reinforced, the HI lines become stronger. Examples: Spica, Rigel.
- A: white stars, $T = 7000/8000$–$10{,}000/15{,}000$ K, $0.0 < B$–$V < 0.3$, maximum intensity of the HI lines and the Balmer series. The lines of CaII are strengthened, those of HeI disappear, and those of FeI appear. Examples: Vega, Sirius.

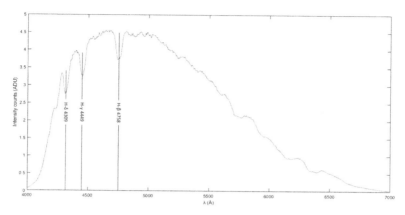

Figure 10.2. A Vega spectrum in arbitrary units. Vega is a A2 star with very clear Balmer series absorption lines. Some absorption lines due to the Earth's atmosphere can also be seen.

- F: yellow–white stars, $T = 6000–7000/8000$ K, $0.3 < B–V < 0.6$, the lines of Caıı become stronger and stronger, the lines of various metals emerge, including Feı and Feıı. Examples: Canopus, Polaris.
- G: yellow stars, $T = 4000/5000–6000$ K, $0.6 < B–V < 1.5$, the Hı lines weaken further, the Caıı lines are very strong, the metal lines are stronger. Examples: Eta Boötis, Sun.
- K: orange–yellow stars, $T = 3000/3500–4000/5000$ K, $1.1 < B–V < 1.5$, the metallic lines dominate the spectrum, the molecular bands begin to be visible. Examples: Arcturus, Aldebaran.
- M: red stars, $T = 2000–3000/3500$ K, $B–V > 1.5$, the Caı line at 422.7 nm becomes very strong and many lines of neutral metals and also of molecular bands like TiO appear. Examples: Antares, Betelgeuse.

In a stellar spectrum we can observe, superimposed, the absorption lines of our atmosphere. The stellar lines have a width linked to thermal agitation, turbulence, rotation (Doppler) and gas pressure (due to interactions with other ions). In addition, the spectral profile is dominated by the spectral response curve of the instrument (Figure 10.2).

10.2 Broadband Photometry

The most direct way to select different frequencies for observation is to use filters. Broadband filters can be made from intrinsically opaque materials at certain wavelengths. They are used to select a band of interest or to protect or reduce the background of the detector. In the millimeter range, materials such as Teflon, polyethylene, Yoshinaga, and Pyrex are used (Figure 10.3).

Figure 10.3. Mesh filter used in the South Pole Telescope[1] at 4 K (left) and at 250 mK (right). Credit: Wikipedia: Lizinvt (CC BY-SA 3.0).

Depending on the wavelength of interest, one must be careful that what is not transmitted through by filter is either reflected (attention has thus to be paid to standing waves) or absorbed, and this therefore determines the emissivity of the filter material. In optical and near-infrared astronomy, standards have been defined (a so-called photometric system). We adopt the following standard:

- U = ultraviolet
- B = blue
- V = visual
- R = red
- I = infrared.

Among the most-used filters in the visible regime are glasses that are suitably treated (i.e., colored) to absorb some wavelengths and transmit others. In order to obtain narrower filters, multiple filters can be used in a cascade.

If we want to have more precise information on the emission frequency of a source, we must use narrower filters. One method is to use double-layer filters (Figure 10.4). Since every time a ray hits a surface it is partially reflected, part of the radiation takes a longer path before escaping from the double layer, and therefore the output radiation will be a combination of many effects. Ideally only one wavelength (and its multiples) will contribute constructively, while the other ones will interfere destructively. In this way, a narrow bandpass filter is fabricated.

In the far-infrared and submillimeter, metallic meshes are used that behave like capacitive or inductive "transmission lines." Usually several meshes are suitably spaced within plastic materials ("hot pressed"; Figure 10.5).

---

[1] https://astro.uchicago.edu/research/spt.php

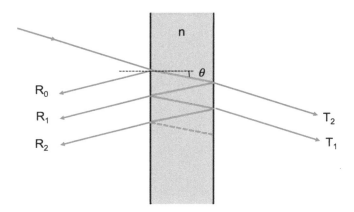

Figure 10.4. Double-layer interference filter.

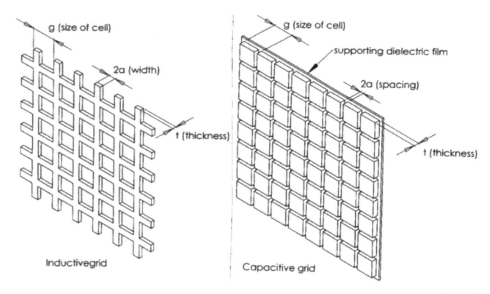

Figure 10.5. Metallic mesh filters. Left: an inductive filter made of a net of metallic material. Right: a capacitive grid, which is the negative of the inductive grid with squares instead of a net. Credit: Wikipedia: Lizinvt (CC BY-SA 3.0).

Worked Example: Anti-reflection Coating
   In some cases, one wants to minimize the reflected electric field that can arise when light passes from one medium to another. This can be done by means of an anti-reflection coating. When an electric field hits a surface we can consider:
   - E_{0i}, *the incident field*
   - E'_{0i}, *the reflected field*
   - E_{0r}, *the refracted field.*

The square of the electric field is clearly proportional to the intensity I of the incoming radiation.

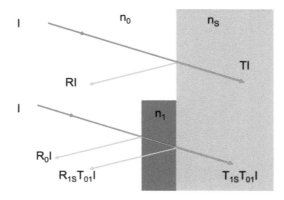

   Imposing the boundary condition we have:

$$E_{0i} = E'_{0i} + E_{or}$$

And from the energy conservation:

$$n_0 E_{0i}^2 = n_0 E'_{0i}{}^2 + n_S E_{or}^2$$

From which, after solving, we have:

$$E'_{0i} = \frac{n_0 - n_S}{n_0 + n_S} E_{oi} \quad E_{0r} = \frac{2n_0}{n_0 + n_S} E_{oi}$$

And thus:

$$R = \frac{E'_{0i}{}^2}{E_{0i}^2} = \left(\frac{n_0 - n_S}{n_0 + n_S}\right)^2$$

Both of the filters and the other optical transmission elements have the difficulty of maximizing the transmission efficiency without losing too much reflected radiation. A solution is to insert a material with an intermediate n, an anti-reflection coating (ARC). To prove this, we do a numerical example:
   Example: $n_0 = 1$, $n_S = 1.5$, $n_1 = 1.3$
   With no ARC:

$$R = \left(\frac{n_0 - n_S}{n_0 + n_S}\right)^2 = \left(\frac{0.5}{2.5}\right)^2 = 0.04$$

With ARC:

$$R_{01} = \left(\frac{n_0 - n_1}{n_0 + n_1}\right)^2 = \left(\frac{0.3}{2.3}\right)^2 = 0.017$$

$$R'_{01} = (1 - R_{01})\left(\frac{n_1 - n_S}{n_1 + n_S}\right)^2 = 0.983\left(\frac{0.2}{2.8}\right)^2 = 0.005$$

And thus

$$R'_{01} + R_{01} < R.$$

10.3 Prism

The refractive index of a medium is not constant but depends on the wavelength. Thus, Snell's law depends on λ:

$$n_2(\lambda_2) \sin\theta_r = n_1(\lambda_1) \sin\theta_i. \tag{10.1}$$

The wavelength dependence of the refractive index can be parameterized using the Hartmann constants (Table 10.1), as in the following formula:

$$n(\lambda) = A + \frac{B}{\lambda - C}. \tag{10.2}$$

If we consider a material with parallel faces, when a beam enters it, the approach to the surface normal (depending on the wavelength) will be balanced by the equal departure from the surface normal when the beam exit the material: incoming white light comes out white. If, however, we consider a material that does not have parallel faces (a prism, Figure 10.6), the approach to the surface normal (depending on the wavelength) can be amplified at the output: we can decompose the light into its components. In 1815, Fraunhofer used a prism to make the first detailed study of the solar spectrum. After him, Father Secchi was among the pioneers of spectroscopy.

The deviation angle of a prism is obtained from Snell's law considering that (see Figure 10.7)

Table 10.1. Hartmann Constants

	A	B	C
Crown glass	1.500	3.5×10^{-8}	-2.5×10^{-7}
Dense flint glass	1.650	2.1×10^{-8}	1.5×10^{-7}
Fluorite	1.429	5.3×10^{-8}	3.6×10^{-7}

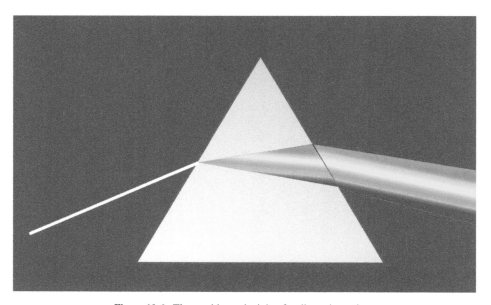

Figure 10.6. The working principle of a dispersion prism.

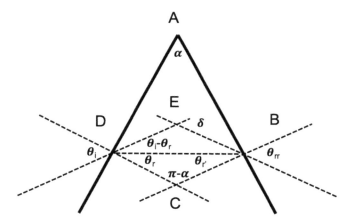

Figure 10.7. Prism deviation angles.

$$\begin{cases} \theta_r + \theta_{r'} = \alpha \\ \theta_i + \theta_{rr} = \delta + \alpha \end{cases}.$$ (10.3)

The angle δ is the global deviation angle, and we must look for the minimum deviation angle $\delta_{\min}$ (at the central wavelength) where $\theta_i = \theta_{rr}$.

$$\begin{cases} \sin(\theta_i) = n(\lambda)\sin(\theta_r) \\ \sin(\theta_{rr}) = n(\lambda)\sin(\theta_{r'}) \end{cases}$$ (10.4)

And so:

$$n(\lambda) = \frac{\sin(\theta_i)}{\sin(\theta_r)} = \frac{\sin[(\alpha + \delta_{min})/2]}{\sin(\alpha/2)}. \tag{10.5}$$

A prism must be used in such a way to reduce aberrations. The light rays from a source are made parallel with a collimator (which is achromatic) and then they pass through a prism that disperses the light depending on its wavelength. Next, they are focused again with an achromatic objective.

The angle of incidence of a beam to a prism must match the minimum deviation angle to reduce aberrations. In the case of an Amici prism, the prism is placed just before the focal plane (with the convergent beam). An objective prism is placed in front of the lens of a telescope (the primary one), but their use is limited by their size. Usually for these applications, the angle of the prism is small (12° maximum).

10.4 Grating

10.4.1 Single Aperture Grating

Let's revisit the slit in Figure 2.15 of Chapter 2, which is reproduced in Figure 10.8. Here, a flat wave front approaches the slit (the aperture), on the opposite side of which is a screen S at infinity. Any point within the aperture, x, can be the origin of a new wave front. If all of the new wave front sources at the aperture are combined as a function of θ, what sort of pattern appears at points P on S?

We must integrate all of the contributions from the points x between 0 and d, assuming a cosine-like incident wave form. For a phase delay of $x\sin(\theta)$, the resulting field intensity will be:

$$I(P) = \frac{a^2 d^2}{2}\left[\frac{\sin^2(\alpha)}{\alpha^2}\right] \tag{10.6}$$

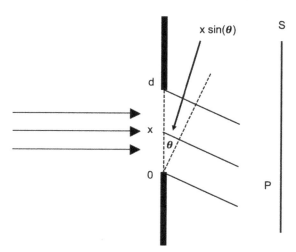

Figure 10.8. Incident electromagnetic wave front to a slit.

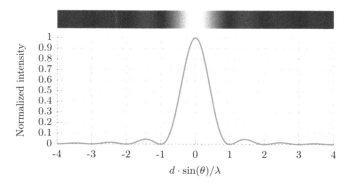

Figure 10.9. Single-slit diffraction pattern.

with:

$$\alpha = \frac{kd \sin(\theta)}{2} = \frac{\pi d}{\lambda} \sin(\theta). \tag{10.7}$$

As a result, we get an interference pattern at S with:
- Constructive interference (light bands) for semi-integers α: $\alpha = \pm \frac{2n+1}{2} \pi$
- Destructive interference (dark bands) for integers α: $\alpha = \pm n\pi$

for $\sin(\theta)$ multiples of λ/d (Figure 10.9). The pattern's central peak has an angular size of:

$$2 \sin(\theta) \cong 2\theta = 2\frac{\lambda}{d}.$$

10.4.2 Double Aperture Grating

Now consider two slits of width d separated by distance D (Figure 10.10):

We will have a double effect, the first linked to the single slit and the second linked to the interference between one slit and the other. The delay between the beams coming from the same position in one slit or the other depends on

$$\delta r = D \sin(\theta). \tag{10.8}$$

So, the phase difference will be

$$\Delta\phi = 2\pi \frac{\delta r}{\lambda} = \frac{2\pi D \sin(\theta)}{\lambda}. \tag{10.9}$$

Adding all of the contributions of I_1 and I_2:

$$I_1(P) = \langle E_1(P, t) \rangle$$
$$I_2(P) = \langle E_2(P, t) \rangle \tag{10.10}$$

we have:

$$I_{\text{tot}}(P) = I_1(P) + I_2(P) + 2\sqrt{I_1(P)I_2(P)}\cos(\Delta\phi) = 2I(P)[1 + \cos(\Delta\phi)]. \tag{10.11}$$

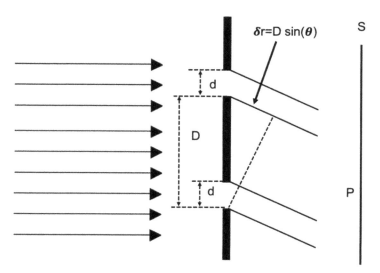

Figure 10.10. Double slit hit by an electromagnetic wave.

Therefore, introducing the expression for $I(P)$; (Equation (10.6)), the total intensity will be the product of two functions

$$I_{\text{tot}}(P) = \frac{a^2 d^2 \sin^2\left(\frac{\pi d \sin(\theta)}{\lambda}\right)}{\left(\frac{\pi d \sin(\theta)}{\lambda}\right)^2}\left[1 + \cos\left(\frac{2\pi D \sin(\theta)}{\lambda}\right)\right]. \tag{10.12}$$

The first cosine wave is due to cross interference with a shorter period (it depends on D). The second function is that seen in single slit interference; it has a long period and acts as an envelope (Figure 10.11). This is exactly what happens in interferometry.

10.4.3 Multiple Aperture Grating

Now we can consider the case of a large number N of slits: a diffraction grating (Figure 10.12). The intensity will have still have an interference pattern shaped like that in Figure 10.11, but the internal modulation will depend on the total length of the slit b and on the number of slits N. There will be primary maxima and much smaller secondary maxima, all enveloped by the single-slit diffraction figure (Figure 10.13). The diffraction figure of a diffraction grating will therefore be:

$$I_{\text{tot}}(P) = \frac{a^2 d^2 \sin^2\left(\frac{\pi d \sin(\theta)}{\lambda}\right)}{\left(\frac{\pi d \sin(\theta)}{\lambda}\right)^2} \frac{\sin^2\left(\frac{\pi b \sin(\theta)}{\lambda}\right)}{\sin^2\left(\frac{\pi b \sin(\theta)}{N\lambda}\right)}. \tag{10.13}$$

The greater the number of slits (N can be large), the narrower the secondary maxima will be.

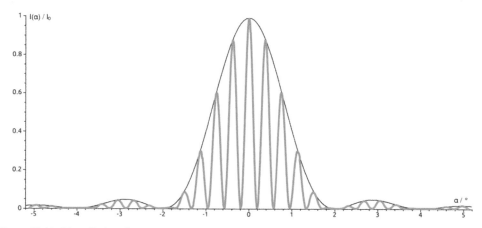

Figure 10.11. Two-slit interference pattern. The contribution of a single slit can be seen in the envelope represented by the black curve. The contribution of the multiple slits is the sine-like curve underneath.

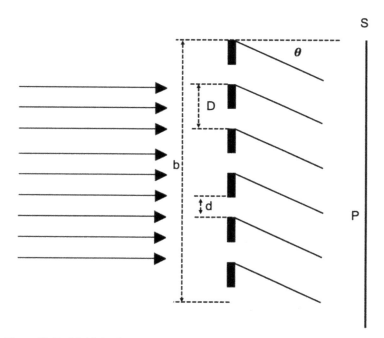

Figure 10.12. Multiple slits (a diffraction grating) hit by an electromagnetic wave.

The primary maxima are very pronounced as N increases and can be far from the origin depending on the grating step size and the order number. We have:

- The main maxima:

$$\sin(\theta) = \pm n\frac{\lambda}{D} \tag{10.14}$$

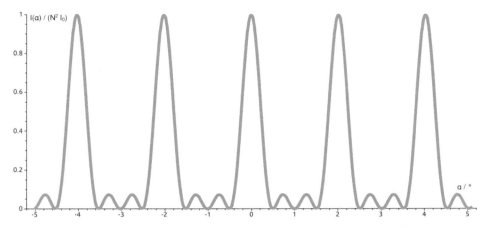

Figure 10.13. Diffraction pattern of a diffraction grating. In addition to the sequence of primary maxima with higher intensity and secondary maxima with lower intensity, the overall curve is modulated by the effect of the single slit, which tends to amplify the primary maxima close to zero and to gradually reduce the maxima of orders higher than zero (not shown in the figure). Credit: Wikipedia: Von Klaus-Dieter Keller—Eigenes Werk, created with SciDAVis Diese W3C-unbestimmte Vektorgrafik wurde mit Inkscape erstellt (CC BY 3.0).

- The secondary maxima:

$$\sin(\theta) = \pm m \frac{\lambda}{ND} \tag{10.15}$$

- Additional secondary maxima caused by the envelope:

$$\sin(\theta) = \pm n' \frac{\lambda}{d}. \tag{10.16}$$

Now, we wonder if this figure is a function of the wavelength. Obviously, the answer is yes.

$$\frac{d\theta}{d\lambda} \rightarrow d \sin(\theta) = \pm n \frac{d\lambda}{D} \rightarrow \cos(\theta) d\theta = \pm n \frac{d\lambda}{D} \rightarrow \frac{d\theta}{d\lambda} = \frac{\pm n}{D \cos(\theta)} \tag{10.17}$$

which means that different wavelengths are focused in different positions. If n increases or D decreases, this dependence is more pronounced.

The spectrographs used in astronomy have from 100 to 3000 slits per millimeter and are several centimeters long, for a total of hundreds of thousands of slits. The zero order ($n = 0$) has no spectroscopic capabilities, but the other orders do. The spectrum will be more dispersed with increasing order n. The different n orders can intersect, so it is necessary to have control of the dispersion (Figure 10.14).

The formulas seen so far work in both transmission and reflection: we just need to obscure (or tilt) different parts of a mirror in order to create a grating. The tiled "ladder" in such a spectrograph is the working principle of a blazed or echelette

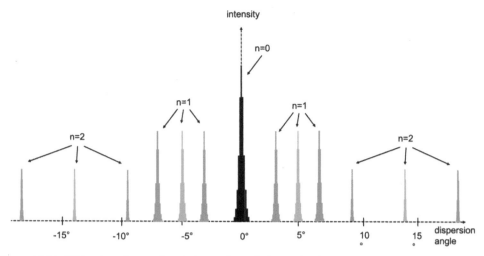

Figure 10.14. Diffraction figure of polychromatic radiation. The first orders $n = 0$ (black, with no spectroscopic capability), $n = 1$, and $n = 2$ are shown.

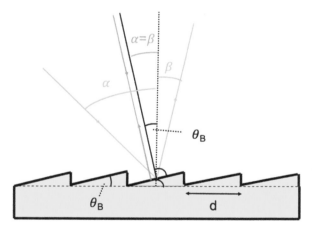

Figure 10.15. A blaze (or echelette) reflection grating.

diffraction grating. One can adjust the blaze angle in order to have a given dispersion direction and to compensate for aberrations. To have a dispersive spectroscopic power, it is necessary, as usual, to look at the maxima of order 1 or greater. A blazed grating is optimized to achieve the maximum efficiency for a given order at a given frequency (Figure 10.15).

The use of a "blazed" grating allows one to align the maximum of the diffraction figure of a single slit (which normally would not have dispersive power) with that of the grating. The inclination of the ladder can be such as to have a reflection angle equal to that which would be obtained with Snell's law; therefore, gratings and prisms are used in the same way. Given that, when using a spectrograph, spectral

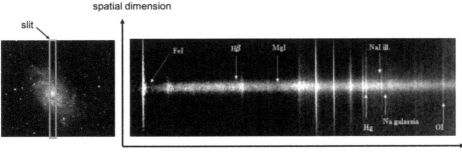

Figure 10.16. A spectrograph image of a galaxy. Left: image on the sky. Right: spectrum along the horizontal direction and the spatial distribution along the vertical direction. The emission lines of different elements such as Fe, Na, and the hydrogen Balmer series are also indicated.

and spatial information are mixed, it is necessary to select the field of view in the dispersion direction: this is the function of the slit. The dispersion occurs in the direction orthogonal to the slit in order to observe several sources simultaneously. On a CCD, in one direction we have the spectrum and in the other we have the spatial distribution (Figure 10.16).

10.4.4 Grism

A grism is a combination of a grating and a prism. The grating can be engraved or glued on the prism (and be of different material). The dispersing element is the grating and the prism has the task of realigning the direction of the output spectrum in the direction of the optical axis. Its use is optimal if the beam is parallel to the focal plane thus if the *f#* is high. There is therefore the advantage of having an on-axis system with reduced off-axis aberrations. In addition, with appropriate precautions, the system works in blaze conditions to maximize efficiency.

10.5 Fourier Transform Spectrometer

The working principle of a Fourier transform spectrometer (FTS) is based on a number of sequential steps:
- A beam of light is emitted by a source (or focused by a telescope).
- This beam is divided by a beam splitter into two equal components.
- The phase of one of the two components is shifted with respect to the other using a movable mirror.
- The beam is recombined at the same beam splitter.
- Depending on the relative positions of the two mirrors (and therefore of the optical path difference), we can have constructive or destructive interference.
- The resulting signal as a function of the optical path difference is called the interferogram and is the Fourier transform of the spectrum of the source.

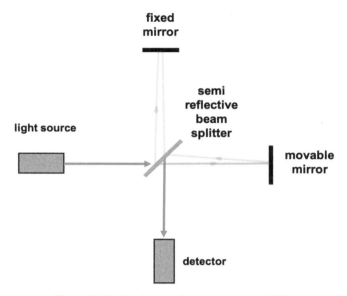

Figure 10.17. Fourier transform spectrometer (FTS).

All FTSs have a moving part (e.g., a moving mirror) and a beam splitter (Figure 10.17). The different kinds of FTSs differ depending on additional components. The most used are:
- Michelson spectrometer
- Martin–Puplet interferometer
- Lamellar grating.

If we consider a monochromatic wave of electric field E_1 and wavelength $\lambda = 1/\sigma$, and the analogous wave of electric field E_2 that is delayed by an optical path difference x, we have:

$$
\begin{aligned}
E_1(t) &= E_0(\sigma)\cos(2\pi\sigma ct) \\
E_2(t) &= E_0(\sigma)\cos(2\pi\sigma ct + 2 \cdot 2\pi\sigma x).
\end{aligned}
\tag{10.18}
$$

If we consider the field intensity as a function of the optical path difference, this is the so-called interferogram $I(r)$:

$$
\begin{aligned}
I(x) &= \left\langle \frac{(E_1(t) + E_2(t))^2}{2} \right\rangle \\
&= \frac{E^2{}_0(\sigma)}{2}[\langle\cos^2(2\pi\sigma ct)\rangle + \langle\cos^2(2\pi\sigma ct + 4\pi\sigma x)\rangle \\
&\quad + \langle 2\cos(2\pi\sigma ct + 4\pi\sigma x)\cos(2\pi\sigma ct)\rangle].
\end{aligned}
\tag{10.19}
$$

Thus:

$$I(x) = \frac{E^2{}_0(\sigma)}{2}\left[\frac{1}{2} + \frac{1}{2} + 2 \cdot \frac{1}{2}\cos(4\pi\sigma x)\right] = \frac{E^2{}_0(\sigma)}{2}[1 + \cos(4\pi\sigma x)]. \quad (10.20)$$

So, for a monochromatic wave the interferogram has a cosinusoid form (plus an offset) because, for every full wavelength, the interference is constructive and, for every half wavelength, the interference is destructive.

The oscillation depends on the wave number and the movement of the retarder: in the position of no delay (zero path difference), the waves add up constructively and then repeat periodically on every wavelength.

If the electromagnetic wave is polychromatic, constructive interference for all frequencies will only occur at the position of zero path difference (zpd), and there will be less intense secondary maxima as the optical path difference increases and decreases (Figure 10.18).

Polychromatic radiation is formed by the sum of many monochromatic waves. In this case, the interferogram can be obtained by adding up all of the contributions.

$$I(x) = \int_0^\infty \frac{E^2{}_0(\sigma)}{2}[1 + \cos(4\pi\sigma x)]d\sigma = \frac{1}{2}I_0 + \frac{1}{2}\int_0^\infty I(\sigma)\cos(4\pi\sigma x)d\sigma \quad (10.21)$$

The interferogram will be given by a constant signal (total power) plus a modulated part that is the cosine Fourier transform of the spectrum. The spectrum will be the Fourier anti-transform of the interferogram. The spectrum and interferogram are Fourier conjugate functions.

Then we have:

$$I'(x) = \int_0^\infty I(\sigma)\cos(4\pi\sigma x)d\sigma \quad (10.22)$$

$$I(\sigma) = \int_{-\infty}^\infty I'(x)\cos(4\pi\sigma x)dx. \quad (10.23)$$

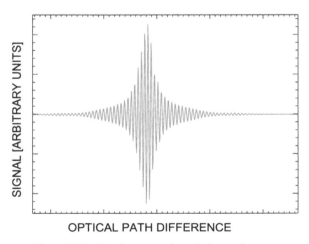

Figure 10.18. Interferogram of a polychromatic wave.

In practice it is not possible to extend the optical path difference to infinity, so from an interferogram we can only get an approximation of the true spectrum. The frequency resolution is then dictated by this maximum optical path difference:

$$\tilde{I}(\sigma) = \int_{-x_{max}}^{x_{max}} I'(x)\cos(4\pi\sigma x)dx \rightarrow \Delta\sigma = \frac{1}{x_{max}} \quad (10.24)$$

That said, the acquisition of an interferogram is often carried out by sampling the signal with non-infinitesimal optical path differences. This, due to aliasing, means that the measured frequencies are limited. Practically we have:

$$\sigma_{max} = \frac{1}{2\Delta x}. \quad (10.25)$$

Chapter 11

Polarization

This chapter is dedicated to polarization. The need for polarization measurements will be described. The definition of Stokes parameters will be given. Polarization in radio astronomy and the way to detect it with correlation polarimeters will be described. Also, polarization at millimeter wavelengths and the use of wire grids, half-wave plates, quarter-wave plates, and Stokes polarimeters are outlined. Lastly, polarization at optical and X-ray wavelengths is discussed.

11.1 The Importance of Polarization Measurements

The study of polarization is important at all wavelengths. The stellar radiation is not very polarized; however, once the emitted radiation interacts with interstellar dust it can acquire a degree of polarization. Astrophysical radiation can also be intrinsically polarized, but a net polarization is observed only in cases when individual parts of a source are aligned or because of diffusion/projection effects. In general, polarization helps to disentangle theoretical models. Polarization can be used to study interstellar magnetic fields and also Zeeman splitting. The Sun emits both linear and circular polarization, although it is more difficult to study polarization in the visible than in other bands. Galactic magnetic fields can be studied through Faraday rotation.

The cosmic microwave background (CMB) is weakly polarized and one can distinguish two different patterns of polarization: E-modes and B-modes. Synchrotron radiation and thermal dust emission are polarized and both constitute emission sources, or a foreground signal, for CMB measurements. The Earth's atmosphere does not usually pose a problem for CMB studies, but one should pay attention to the Earth's magnetic field, which produces Zeeman splitting and, consequently, circular polarization (Figure 11.1).

doi:10.1088/2514-3433/ac0ce4ch11   11-1

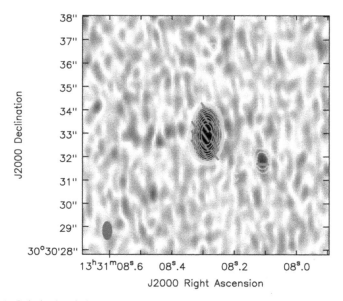

Figure 11.1. Polarized emission of the quasar 3C286. Credit: Wikipedia: ALMA (CC BY 4.0).

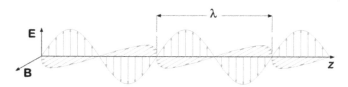

Figure 11.2. A linearly polarized electromagnetic wave. Credit: Wikipedia: P.wormer (CC BY-SA 3.0).

11.2 Stokes Parameters

The electric field E (as well as the magnetic field B) can be decomposed, at any instant, into two mutually orthogonal components directed along an x-axis and a y-axis:

$$E_1 = \hat{i} E_{01} \cos(kz - \omega t + \varphi_1)$$
$$E_1 = \hat{j} E_{02} \cos(kz - \omega t + \varphi_2). \tag{11.1}$$

If the phase difference $\varphi_1 - \varphi_2$ between them varies randomly over time and space, then E changes orientation and plane of oscillation randomly over time. The wave is said to be non-polarized.

If, however, the phase difference remains constant, the electric field has a defined oscillation plane and the light is said to be polarized. If the phase difference is 0 or π, then the electric field vector oscillates in the same plane and the polarization is linear (Figure 11.2). If the phase difference is $\pi/2$ there is an elliptical polarization (which is circular if $E_{01} = E_{02}$), and the ellipse is rotated in the other cases (Figure 11.3).

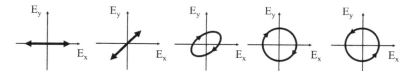

Figure 11.3. A polarized wave. From left to right: a linearly polarized wave along the x direction; a linearly polarized wave oriented at 45°; an elliptically polarized wave with a phase difference of 7/4 π; a right circular polarized wave; a left circular polarized wave.

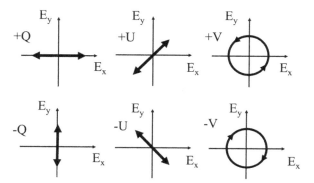

Figure 11.4. Stokes parameters. Left: 100% Q, positive (top), and negative (bottom). Center: 100% U, positive (top), and negative (bottom). Right: 100% V, positive (top), and negative (bottom).

An electromagnetic wave can be totally or partially polarized. In the second case, we define the degree of polarization P $(0 \leqslant P \leqslant 1)$ as in the following:

$$P = \frac{I_{\text{pol}}}{I_{\text{pol}} + I_{\text{unpol}}}. \tag{11.2}$$

The Stokes parameters I, Q, U, and V describe the polarization state (even partial) of light. Q and U describe the amount of linear polarization, and V describes the circular polarization. The amount of total polarization of a wave can be defined as:

$$P^2 = Q^2 + U^2 + V^2. \tag{11.3}$$

The Stokes parameters are defined as:

$$I = E_{01}^2 + E_{02}^2 \tag{11.4}$$

$$Q = E_{01}^2 - E_{02}^2 \tag{11.5}$$

$$U = 2E_{01} E_{02} \cos(\varphi_1 - \varphi_2) \tag{11.6}$$

$$V = 2E_{01} E_{02} \sin(\varphi_1 - \varphi_2). \tag{11.7}$$

I is the total power of the wave. Q indicates if there is polarization along the x- (in this case it is positive) or y- (in this case it is negative) axes. U is also an estimation of the linear polarization, but at +/− 45°. V gives information about direction of the circular polarization. The use of the Stokes parameters can be seen in Figure 11.4.

11.3 Polarization at Radio Frequencies

A radiometer is based on the operating principle of an electric dipole or an antenna, so it is intrinsically sensitive to one polarization. There are single polarization or double polarization radio receivers. The latter are sensitive to the two linear polarizations or to the two circular polarizations. With appropriate changes of coordinates, one can derive linear polarization from a circular one and vice versa (Figure 11.5).

For the linear "natives" we have:

$$\begin{cases} E_x^2 = XX \\ E_y^2 = YY \end{cases} \rightarrow \begin{cases} I = XX + YY \\ Q = XX - YY \end{cases}. \tag{11.8}$$

For circular natives (for which a 90° delay is added to one of the two components) we have:

$$\begin{cases} LL \\ RR \end{cases} \rightarrow \begin{cases} I = LL + RR \\ V = LL - RR \end{cases}. \tag{11.9}$$

The ability of radiometers to detect radiation in a coherent way can again be exploited. The two polarizations XX and YY divide. If a heterodyne system is used, the same local oscillator (LO), possibly out of phase, can be used to build a correlation radiometer. The two components (IF) are multiplied by means of a multiplier, and a second multiplication is carried out by introducing a 90° phase shift between the two. In this way we get four outputs z_1, z_2, z_3, and z_4 from which we get (Figure 11.6):

$$\begin{cases} I \propto z_1 + z_2 \\ Q \propto z_1 - z_2 \\ U \propto z_3 \\ V \propto z_4 \end{cases}. \tag{11.10}$$

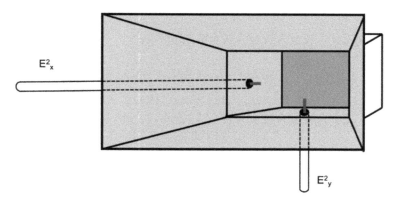

Figure 11.5. A double polarization radiometer antenna.

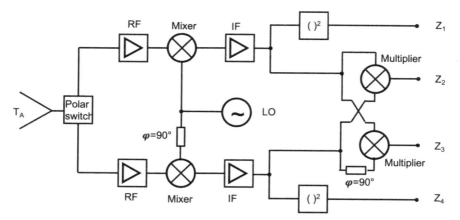

Figure 11.6. A correlation polarimeter.

11.4 Polarization at Millimeter and Submillimeter Wavelengths

Between the millimeter and the far-infrared bands, electromagnetic signals are detected incoherently. In order to detect polarization, quasi-optical methodologies have to be used: a signal is divided into two orthogonal polarizations, one of the two is set out of phase, and the two parts are recombined. Then, we study how the two polarizations interact. To divide the two orthogonal polarizations, quasi-optical elements are used that have properties that depend on polarization. A wire grid is a polarizer that converts a non-polarized wave into a totally polarized one (Figure 11.7).

A wire grid is typically composed of tungsten wires whose diameter is much smaller than the spacing between them, which in turn is much smaller than λ. An electromagnetic wave oscillating in the direction of the tungsten wires "sees" the quasi-optical element as if it were a mirror and it is reflected back. An orthogonal polarized electromagnetic wave is, on the other hand, transmitted by the wire grid. Non-polarized radiation, or radiation polarized at 45° with respect to the wires in the grid, is thus split in two components, one reflected and one transmitted, orthogonally polarized. In a polarimeter, in order to select one polarization or another, the optical elements active in the modulation are rotated.

Partially polarized incoming radiation is transmitted completely if the polarization is orthogonal to the wires. If the wire grid rotates with an angular velocity ω, then we will have a signal with an intensity W:

- If the light is not polarized,

$$W = \frac{1}{2}I.$$ (11.11)

- If the light is totally linearly polarized with $I = P$,

$$W = \frac{P}{2} \cdot \cos(2\omega t) + \frac{P}{2}.$$ (11.12)

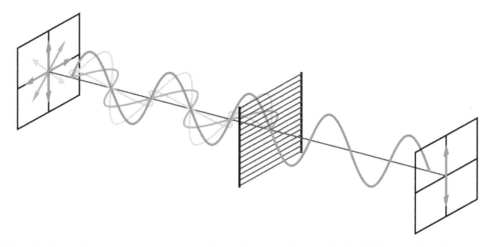

Figure 11.7. An electromagnetic wave through a wire-grid polarizer. Credit: Wikipedia: (CC BY-SA 3.0) https://en.wikipedia.org/wiki/Polarizer.

- If the light is partially polarized with $P < I$,

$$W = \frac{1}{2} \ (I - P) + \frac{P}{2} \cdot \cos(2\omega t) + \frac{P}{2}.$$ (11.13)

The wire grid has no effect on circular polarization.

Another optical element that produces a rotation of the polarization vector is a half-wave plate (HWP). This is an optical element built with birefringent material. Birefringent materials have a different refractive index in the two orthogonal axes (ordinary n_o and extraordinary n_e); therefore, they cause a phase shift:

$$\Delta\varphi = \frac{2\pi d}{\lambda}(n_e - n_o).$$ (11.14)

If constructed of the right thickness, an HWP can introduce a delay on one linear polarization and not on the other one, thereby modifying or rotating the input polarization. In the case of an HWP, the delay is a half wave: $\Delta\varphi = 180°$ (Figure 11.8). If the HWP is rotating, the relative angle between the polarization and the axis of the HWP vary over time. An input linear polarization is rotated by the HWP at 4ω: two times due to the HWP, and two times due to the fact that the polarization vector does not have a versus so an angle $\varphi = \varphi + 180°$.

If we put a wire grid in series with an HWP, then we have built a Stokes polarimeter: a system that would modulate at 4ω only the linearly polarized part of incoming radiation, leaving the non-polarized part unaltered.

Circular polarization is a little more difficult to study. In this case, the relative angle between the polarization and the axis of the HWP vary over time. A quarter-wave plate (QWP) is a quasi-optical system similar to an HWP, but with half the depth. This means that the phase of the radiation is shifted by 90°. In this way, a

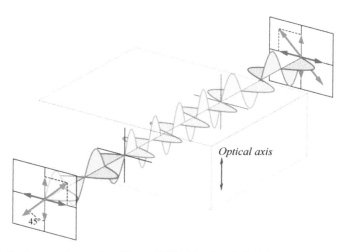

Figure 11.8. Polarization rotation produced by an HWP. The electric field is the composition of components parallel to the *x*- and *y*-axes and is directed at 45°. It is rotated by the half-wave plate by delaying one component with respect to the other. Credit: Wikipedia: (CC BY-SA 3.0) https://commons.wikimedia.org/w/index.php?curid=760346.

circular polarization is converted into a rotating linear polarization. If a wire grid is placed in series with a QWP, a 2ω modulation is obtained.

11.5 Optical Polarization

In the visible, optical elements that absorb polarization in one direction and transmit in another are mainly used. Dichroic crystals absorb one polarization more than another (e.g., tourmaline). Birefringent materials deflect one polarization more than another (e.g., quartzes, sapphires, polymers, etc.). Polaroid is made by immersing needle-like crystals in a polymer (Figure 11.9). During the manufacturing process, the crystals are aligned along a preferential direction by means of a magnetic field, which causes the polarized light to be absorbed along the same direction.

A Wollaston prism is a double prism composed of birefringent material (e.g., calcite) with orthogonally oriented axes (Figure 11.10). On the contact surface between the prisms it refracts the two polarizations differently.

11.6 X-Ray Polarization

The materials used to detect polarization in the millimeter to optical wavelength range cannot be used to detect X-ray polarization, because the photon energies and their penetrating power are too high. Therefore, other phenomena must be used:

- Bragg diffraction: For wavelengths shorter than a few nanometers, an electromagnetic wave diffused by a crystal follows Bragg's law, according to which it has a diffraction pattern and responds to a single polarization. By rotating the crystal, a modulation is obtained that depends on the polarization itself.

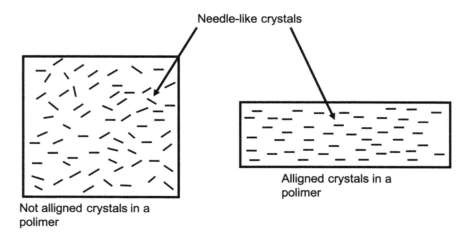

Figure 11.9. Polaroid.

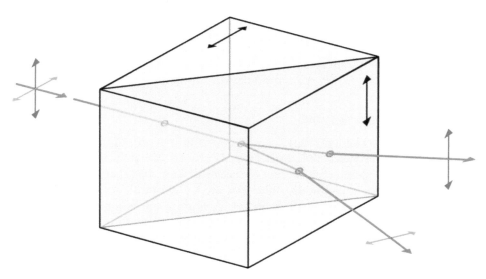

Figure 11.10. Wollaston prism. Credit: Wikipedia: Fgalore (CC BY-SA 3.0).

- Diffusion polarimetry: Compton and Thomson scatterings depend on the polarization. An incident X-ray will produce scattered photons in all directions, but with a prevalence in the direction of the plane of oscillation of the electric field, i.e., polarization.
- Photoelectric effect tracking: Soft X-ray photons can produce a photoelectric effect in a medium. The direction of the photo-produced electron depends on the polarization of the incident light and is preferentially aligned with the incident field, so the measurement of the emission direction of the photo-produced electron is a measure of polarization.

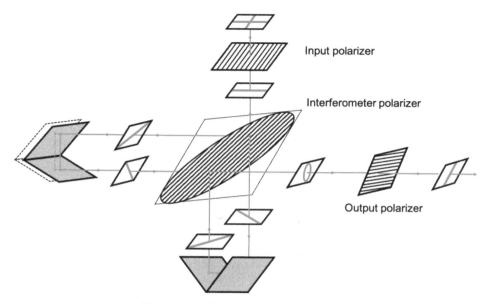

Figure 11.11. Martin–Pupplet interferometer.

11.7 Martin–Pupplet Interferometer

A Martin–Pupplet (MP) interferometer is a Fourier transform interferometer that is based on polarization, particularly on the use of wire-grid polarizers (Figure 11.11). We have three wire grids: an input wire grid, a central wire grid, and an output wire grid. In an MP, the incoming radiation must be polarized at 45° with respect to the central wire grid, so the input wire grid is positioned at 45° with respect to the central wire grid.

Radiation then undergoes beam (polarization) splitting, which introduces a phase delay in one component. Half of the input radiation is reflected by the central wire grid and the other half is transmitted. The radiation is then reflected by roof mirrors, which reflect back the beam with the phase delay. The roof mirrors also flip the polarization by 90°. Thus, the part that was reflected is now transmitted by the central wire grid, and vice versa. The radiation is now recombined with a path difference, which produces an interferogram. In order to exploit the interferogram itself, we need a third wire grid that selects the right polarization and ensures the modulation of the radiation.

Experimental Astrophysics

Elia Stefano Battistelli

Chapter 12

Signal Extraction from Noise and Calibration

Calibrations and the concept of responsivity and sensitivity will be introduced here. System temperature for radiometers and noise equivalent power for millimeter detectors will also be addressed. In addition, this chapter will cover CCD data conditioning, including bias, dark, and flat monitoring, the removal of cosmic rays, photometric calibration, and absolute calibration, along with the use blackbodies for calibration. The concept of a signal-to-noise ratio is also re-presented here. Filters, amplifiers, and lock-in amplifiers will be described. Differential photometry is also introduced and given as a worked example.

12.1 Signal Extraction from Noise: Filters and Amplifiers

In order to detect a signal immersed in noise, a first option is to use a filter. If we want to detect AC signals, we can insert high-pass (RC) filters, trusting on the fact that a capacitor only lets high frequencies pass. An example of a high-pass filter can be seen in Figure 12.1.

If, however, we want to detect a slowly varying signal, we can use a low-pass filter (Figure 12.2).

Combining a high-pass filter and a low-pass filter with different cut-off frequencies allows one to make a bandpass filter (Figure 12.3).

In Fourier space, these filters allow one to reject (filter out) some frequencies and let pass others. An astrophysical signal is first detected by a detector, transduced, possibly filtered, and then amplified. An astrophysical signal is usually a weak signal that must be amplified, taking care not to add noise due to the amplification system itself. Of course, if the detector (the transducer) has a high level of intrinsic noise, the amplifier will also amplify that and there is no gain in terms of the signal-to-noise ratio (S/N). That said, if the detector noise is low, we should be careful that the use of an amplifier or a filter does not degrade the signal. In electronics, amplifiers are based on the use of transistors (e.g., a p–n–p junction). Integrated amplifiers (operational amplifiers) are a collection of many transistors on a single chip (Figure 12.4).

doi:10.1088/2514-3433/ac0ce4ch12

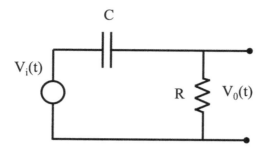

Figure 12.1. High-pass filter.

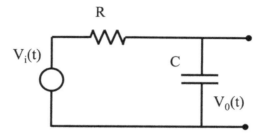

Figure 12.2. Low-pass filter.

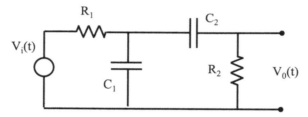

Figure 12.3. Bandpass filter.

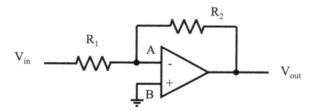

Figure 12.4. A differential operational amplifier.

12.2 Modulation

A DC amplifier often has significant problems related to stability, gain, and offset. It is usually affected by $1/f$ noise issues. Even when it is possible to reduce the noise (during construction) of an amplifier such that it is negligible, the detector noise becomes important. In order to reduce the effects of amplified noise, one method that can help is that of modulation, in which the input signal is periodically interrupted, transforming it from a DC to an AC square wave. In this way a differential measurement is made between the "negative" signal and the "positive" signal: the variation of this difference is then the signal. If, with an amplifier, we can tune (filter) on that modulation frequency, we can increase S/N because we can only amplify the modulation frequency with a very small frequency bandwidth around it.

Modulation can take place, for example, by interrupting the emission of the source through its power supply, or by placing an obstruction between the source and the detector through a "chopper," a mechanical blade that rotates in front of a source. Since we introduced the modulation artificially, we can perfectly know the modulation frequency of the signal and filter around it. This allows us to "decide" the amplification frequency, for example, away from the $1/f$ knee (Figure 12.5).

Modulation should be applied as early (i.e., as close to the signal source) as possible, ideally by turning the source off and on because demodulation is effective on the noise that comes after. An example of a measurement and reading chain is

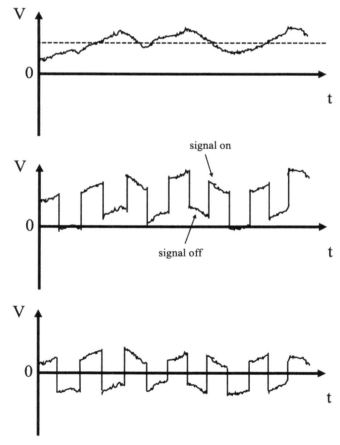

Figure 12.5. The working principle of modulation: a DC signal affected by $1/f$ noise (top) is modulated between an on and an off state (center). The result is shown in the bottom figure.

shown in the following paragraph. As long as we perfectly know the filtering frequency f_0, the higher the quality factor of the filter, and the higher the S/N of the retrieved signal.

12.3 Lock-in Amplifiers

If the modulation is carried out mechanically, or with instrumentation whose characteristics vary over time, the modulation frequency can vary and therefore we would need to use a filter with an adequate width and, ultimately, acquire more noise. An alternative method (in reality, it is an additional one) is that of synchronous demodulation. A "chopper" is not only a mechanical blade with which to block the radiation, it also produces a reference signal that is perfectly synchronous with the modulation (Figure 12.6).

For example, an LED + photodiode system mounted on the wheel could give an exactly synchronous signal (in phase) with the modulation of the source. A synchronous (locked-in) demodulator is an apparatus that allows one to extract a (modulated) signal from a signal dominated by noise. The advantage of the synchronous demodulator (compared to that of a simple filter) is that of knowing both the modulation frequency and the phase: with a reference signal generated by the same modulation apparatus, it is not only known at what frequency the wave front is blocked, but when. The lock-in amplifier "locks" the signal to a reference, so it has two inputs (Figure 12.7):

- One for the signal, which will also contain noise $s(t) + n(t)$
- One for the reference $r(t)$.

The lock-in amplifier measures the quantity:

$$v_{\mathrm{p}}(t) = r(t) \cdot [s(t) + n(t)]. \tag{12.1}$$

And it makes a temporal average on a time constant τ that is suitably long:

$$v_{\mathrm{out}}(t) = \left\langle v_{\mathrm{p}}(t) \right\rangle_{\tau} = \left\langle r(t) \cdot [s(t) + n(t)] \right\rangle_{\tau}. \tag{12.2}$$

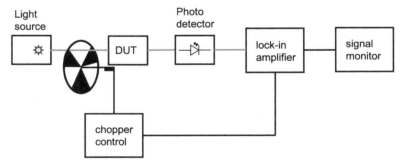

Figure 12.6. Signal processing with a chopper. A light-source signal goes through a chopper, a mechanical blade that blocks the source periodically. Light crosses a device under test (DUT) and then is detected. The signal is demodulated by use of a lock-in amplifier.

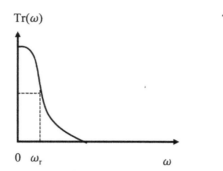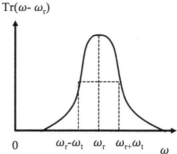

Figure 12.7. Lock-in filtering.

The signal can be broken down as follows:

$$v_{\text{out}}(t) = \langle r(t) \cdot [s(t) + n(t)]\rangle_\tau = \langle r(t) \cdot s(t)\rangle_\tau + \langle r(t) \cdot n(t)\rangle_\tau. \qquad (12.3)$$

And since the signals $s(t)$ and $r(t)$ are correlated, the time average (the first term) will produce a constant positive result. The signals $n(t)$ and $r(t)$ are instead uncorrelated; therefore, the time average (the second term) will produce a signal that fluctuates around zero and therefore will be zero mean. The average over the time constant τ will not allow for the detection of variations in the signal that are faster than τ, so τ must be long enough such that the median noise value is zero, but short enough to cause the signal to change more slowly than τ.

Let's assume only cosinusoidal signals at different frequencies for the time being:

$$s(t) = \sqrt{2}\, V_s \cos(\omega_s t + \phi_s) \qquad (12.4)$$

$$r(t) = \sqrt{2}\, V_r \cos(\omega_r t + \phi_r). \qquad (12.5)$$

From the properties of the sums and differences of cosines, we have that the lock-in filter produces:

$$
\begin{aligned}
v_p(t) &= s(t) \cdot r(t) \\
&= V_r V_s \cos\big[(\omega_s + \omega_r)t + \phi_s + \phi_r\big] \\
&\quad + V_r V_s \cos\big[(\omega_s - \omega_r)t + \phi_s - \phi_r\big]
\end{aligned} \qquad (12.6)
$$

which has one term at the sum frequency and one term at the difference frequency. If we now apply a low-pass filter with an intermediate cut-off frequency ω_{cut} such as

$$\omega_s - \omega_r < \omega_{\text{cut}} \ll \omega_s + \omega_r \qquad (12.7)$$

only the signal at the difference frequency will survive

$$v_p(t) = V_r V_s \cos\big[(\omega_s - \omega_r)t + \phi_s - \phi_r\big] \qquad (12.8)$$

that contains in itself the information about V_s and removes the information at frequencies different from ω_r. And since removing all frequencies far from a fixed frequency means making a bandpass, in fact, filtering with a low-pass filter with a width ω_t after modulating and multiplying the two signals is equivalent to making a bandpass around ω_r with a width of $2 \times \omega_t$.

If we then set the frequency of the signal equal to the reference frequency, $\omega_s = \omega_r$, we can make a filter with a very narrow cut-off frequency, and therefore (if the filter has a gain of 1) the output signal from the lock-in amplifier is a continuous signal whose amplitude depends on the amplitude of the input signal and the difference in phases between the signal and reference:

$$v_p(t) = V_r V_s \cos(\phi_s - \phi_r). \tag{12.9}$$

The difference between the phases in a lock-in amplifier can usually be adjusted. The noise present in the output signal from the lock-in amplifier is only that in the band adjacent to ω_r. In practice, in analog electronics, it is not easy to multiply two signals. However, there are multiplier circuits that produce different distortions and anyway cannot be adjusted. Alternatively, multiplication can be done by converting the signals to digital through an analog-to-digital conversion with a high resolution, low noise, and a high speed, but this can be a very expensive option. Instead, a phase sensitive detector can be used in which the multiplier is replaced by a signal deviator that changes the amplification of a signal between 1 and −1 depending on whether a reference signal is positive or negative. The two are then multiplied (Figure 12.8).

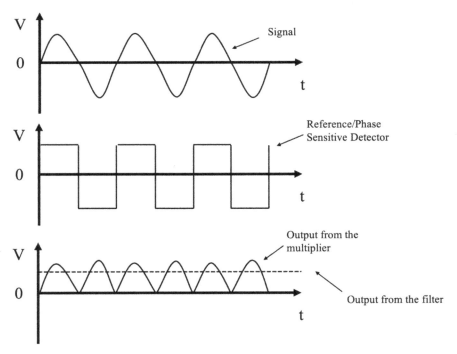

Figure 12.8. The signal (top), a signal deviator (center), and the two multiplied (bottom).

A square-wave signal at amplitude 1 can be broken down into a Fourier series as:

$$r(t) = \frac{4}{\pi} \left\{ \cos\left[\omega_r t + \phi_r\right] - \frac{1}{3}\cos\left[3 \cdot (\omega_r t + \phi_r)\right] \right.$$
$$\left. - \frac{1}{5}\cos\left[5 \cdot (\omega_r t + \phi_r)\right] - \cdots \right\}. \qquad (12.10)$$

Then,

$$v_p(t) = \frac{2 \cdot \sqrt{2} \cdot V_s}{\pi} \left\{ \cos\left[(\omega_r - \omega_s)t + \phi_r - \phi_s\right] \right.$$
$$- \frac{1}{3}\cos\left[(3 \cdot \omega_r - \omega_s)t + 3 \cdot \phi_r - \phi_s\right] \qquad (12.11)$$
$$\left. - \frac{1}{5}\cos\left[(5 \cdot \omega_r - \omega_s)t + 5 \cdot \phi_r - \phi_s\right] - \cdots \right\}.$$

As a result, in addition to a signal at $\omega = 0$, we could also find signals from the higher harmonics that add noise.

If a signal V_s is affected by $1/f$ noise and white noise with a power spectrum equal to $w_n(f)$, the output signal is:

$$S = V_{out} = \langle V_p \rangle_\tau = V_r V_s \cos(\phi_s - \phi_r) = a V_s. \qquad (12.12)$$

So, if we define the filter bandwidth to equal $1/\tau$, the noise will be integrated within the bandwidth:

$$N = \sqrt{\langle \Delta V_{out}^2 \rangle} = \sqrt{a^2 \int_0^{\frac{1}{\tau}} w_n df} = \sqrt{a^2 \frac{w_n}{\tau}} \qquad (12.13)$$

from which:

$$\frac{S}{N} = \frac{V_{out}}{\sqrt{\langle \Delta V_{out}^2 \rangle}} = V_s \sqrt{\frac{\tau}{w_n}} \qquad (12.14)$$

which is proportional to τ (of course, assuming that all of the signal is detected). This is a useful formula for estimating integration times and for measuring where noise is minimal.

12.4 Modulations in Astrophysics

The difference created during modulation can be done either optically or by a "noise" source. In the following scheme, a chopper lets the radiation pass through from the sky when it is open and it reflects the emission of a blackbody when it is closed. All of the emission sources in the path after the chopper are present in both signals and are therefore canceled out in the difference (e.g., filters, mirrors, optics, a window, etc.). However, if the blackbody (and the chopper) are at room temperature, there is a

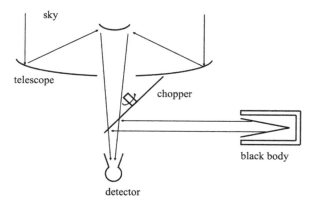

Figure 12.9. Acquisition system with a source, a chopper, a receiver, and a blackbody acting as a reference.

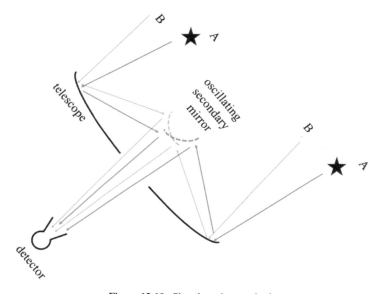

Figure 12.10. Sky-chopping method.

problem with dynamics that has to be sorted out (Figure 12.9). Furthermore, the emission of the atmosphere is not removed.

The atmosphere affects observations in the infrared and radio, and its fluctuations induce noise that is sometimes orders of magnitude stronger than the astrophysical signals. We could try, using synchronous modulation and demodulation techniques, to remove atmospheric fluctuations. However, as mentioned, modulation should take place as soon as possible, because it reduces the noise of everything after it. For these reasons, an angular modulation of the source is performed in the sky. An optical element is inserted in the telescope that allows an astronomer to observe two adjacent sky regions and in fact measure the difference between two fields in the sky (sky chopping; Figure 12.10). For example, we can swing the secondary mirror of a two-mirror telescope. The same system that produces the oscillation produces a reference signal that can be sent to a lock-in amplifier. The signal "seen" by the detectors is alternatively the signal from a source and the signal in a reference field B. The emission signal of the atmosphere and its slow fluctuations are eliminated by the lock-in amplifier.

12.5 CCD Calibration

A CCD camera usually saves the data in arbitrary unit counts on a map (counts versus x and versus y). The raw image is a matrix of positive numbers that is then processed through calibration procedures. In the counts there is the image from the sky and the noise (generated by the detector, the dark current, the read-out, and the environment). We will therefore have to carry out four or five fundamental operations in order to correctly interpret this image. These calibration steps include an operation between matrices that have to be always normalized for their integration time. This normalization can be done in several ways. The easiest way is to divide each acquisition frame with a non-zero exposure time by its exposure time. We denote the normalized frame with a subscript t: $<frame>_t$. The calibration operations are the following:

- Bias subtraction
- Dark subtraction
- Flat-field correction
- Cosmic-ray correction
- Photometric calibration.

Bias: If we make an observation with the CCD shutter closed, for 0 s, we get a non-zero signal that comes from the bias voltage, from the reading, and from its fluctuations (Figure 12.11). This signal is equivalent to a "null" signal. If we make a histogram of the readings, we get a Gaussian distribution that reflects this zero signal.

Many such images need to be taken, and the median calculated and subtracted from the final frames. Replacing the median with the average removes from this calculation values far from the Gaussian distribution. Normally this signal should not have a temperature dependence, although a weak dependence can occur because of a secondary effect, such as from the cool-down electronics. The resulting image is the bias median: $<bias>$.

Dark: Depending on the temperature of the CCD, during the time of an exposure, electrons will be released not only by photo-production but also by the effect of temperature. The greater the exposure time, the greater the number of electrons released: this is the dark current. If we do an exposure with the same settings as the targeted observation but with the CCD shutter closed, we can measure this effect. Also in this case, the median of several dark frames can better correct for the dark current than can the average. This temperature-dependent effect is reduced by cooling the CCD. It should be noted that the dark current depends on the exposure time but the bias does not. If we want to remove dark current from an image, we first

Figure 12.11. Bias frame. Credit: Wikipedia: Rawastrodata (CC BY-SA 3.0).

have to remove the bias and then calculate the proportion of dark current present in the image. If, for example, the target exposure time is 3 min, and the dark frame exposure time is 2 min, we can calculate the dark frame $\mathrm{dark}_{3'}^{b}$ cleaned from the bias, and rescaled from the exposure time, as in the following:

$$\mathrm{dark}_{3'}^{b} = (\mathrm{dark}_{2'} - <\mathrm{bias}>) \cdot \frac{3}{2}. \qquad (12.15)$$

Finally, we have to remove the dark image from the target image and clean and rescale the target image for the exposure time $\mathrm{astro\_frame}_{t}^{d}$. We have:

$$\mathrm{astro\_frame}_{t}^{d} = \mathrm{astro\_frame}_{t} - <\mathrm{bias}> - \mathrm{dark}_{t}^{b}. \qquad (12.16)$$

Flat: Each pixel of a CCD is different from the others in terms of gain and sensitivity. In addition, even a perfectly uniform source would generate an uneven signal because different portions of the CCD experience a different optical efficiency. The object image has to be corrected for this effect. This is a multiplicative effect, so it must be reduced by dividing the object image with a flat image. By illuminating the CCD with a flat field (from a light source), for every filter, this non-uniformity is measured. This can be done with different techniques:

- An internal field: using an internal lamp in the telescope dome.
- A dome field: closing the dome of the telescope.
- A twilight flat: taking an image of the sky at sunset or sunrise in the direction opposite to that of the Sun.
- A sky flat: taking an image of the night sky with the telescope out of focus or pointed to a part of the sky without (many) sources.

In the last two cases, multiple images must be acquired to eliminate any star present in the field by taking the median of all of the images. Another technique is that of the *autoflat*. The same astronomical images can be used by averaging many (>10) different images so that the stars do not fall on the same pixels. The highest value for each pixel is discarded from this average in order to remove the contribution of celestial sources. A way to apply the aforementioned technique is to take the median instead of the average of the flat field images. Usually, a flat image is a short image, so only the bias must be subtracted. Nevertheless, we include in the following the complete formula:

$$\text{Flat}_t^d = \text{Flat}_t - <\text{bias}> - \text{dark}_t^b. \tag{12.17}$$

Finally, the frames cleaned of the bias, dark, and flat, astro_frame$_t^c$, can be calculated as:

$$\text{astro\_frame}_t^c = \frac{\text{astro\_frame}_t^d}{\text{Flat}_t^d}. \tag{12.18}$$

Cosmic rays: Cosmic rays can also affect the images. They create spurious events on the CCD unrelated to the position of astronomical objects in the sky. This occurs especially on high mountains. By taking repeated observations of the same celestial field, they will occur only in a subsample of the images and can therefore be eliminated. Pixels with signals that deviate, for example, more than 3 sigma from the average value can be eliminated as random events: a mask is created and subtracted. Clearly, cosmic rays are have more of an impact on satellite observations than on those Earth-based.

Astronomical calibration: It is necessary to convert the measurements expressed in arbitrary units, AUs, to units of flux, brightness, or magnitude. Astronomical calibration allows you to do this step. We need to observe at least one source of known magnitude. We remind the reader that the apparent magnitude of Vega, in all bands, is $m = 0$. Given the following relationship:

$$m = -2.5 \cdot \log(F) + C \tag{12.19}$$

calibration allows one to determine the constant C and go from instrumental magnitude, $-2.5 \cdot \log(F)$, to apparent magnitude m. In order to be consistent between different images, we always divide the counts that give the instrumental magnitude by the exposure time. The calibration is done for measures obtained from images (for every filter) of Vega ($m = 0$) or of other stars of known magnitude. Catalogs of so-called "standard stars" are available for this purpose.[1]

The magnitude B (for example; the same applies for the other bands) of an astronomical source is:

$$B = B_0 + b - k_B X_B + c(B - V) \tag{12.20}$$

where:
- B is the apparent magnitude.
- B_0 is a multiplying coefficient (as this a logarithmic relation, it is summed).

---

[1] http://www.cfht.hawaii.edu/ObsInfo/Standards/Landolt/

- b is the instrumental magnitude.
- k_B is a coefficient that includes the atmospheric extinction in the observed band.
- $X_B = 1/\cos(z)$ is the airmass.
- c_B is a color coefficient caused by the fact that the filters are not ideal and identical to the standard reference. This term is usually negligible.

The idea behind the calibration is to acquire images at different airmasses and construct a straight-line relation of $y = mx + q$ with:
- $y = B - b$
- $x = X_B$
- $m = k_B$ (the variable to be fitted for)
- $q = B_0 + c_B(B - V)$; (the variable to be fitted for).

Once the slope and the intercepts are determined from observations of a known star, the same B_0 and c_B can be applied to other stars.

Experimental Astrophysics

Elia Stefano Battistelli

Appendix A

Atmospheric Physics for Astronomers

A.1 Atmospheric Constituents

The Earth's atmosphere is composed of the following gases:

- Nitrogen (N_2): ~78%
- Oxygen (O_2): ~21%
- Argon (Ar): ~1%
- Water vapor (H_2O): ~0%–6%
- Carbon dioxide (CO_2), neon (Ne), helium (He), methane (CH_4), hydrogen (H_2), krypton (Kr), xenon (Xe), ozone (O_3), nitrogen oxides (NO, NO_2, N_2O_2), carbon monoxide (CO), ammonia (NH_3), sulfur dioxide (SO_2), hydrogen sulfide (H_2S), powders, and aerosols.

The atmosphere can be divided into layers. They are named:

- The troposphere (0–12 km): all H_2O is concentrated here (and atmospheric phenomena originates here).
- The stratosphere (12–50 km): here, temperature increases with height.
- The mesosphere (50–80 km): meteors burn here.
- The ionosphere (60–450 km): this layer is ionized by cosmic rays and filters X-rays.
- The thermosphere (80–700 km): here, temperature increases with height again.
- The exosphere (700–10,000 km): He and H_2 are concentrated here.

While pressure and density decrease exponentially with altitude, the temperature undergoes a reversal due to ultraviolet radiation and the ozone (in the stratosphere; Figure A.1).

The knowledge of the constituents of the atmosphere gives us indications about its physical behavior. For example, the ozone layer in the stratosphere filters UV radiation. The absorption of terrestrial radiation by some gases leads to the

doi:10.1088/2514-3433/ac0ce4ch13

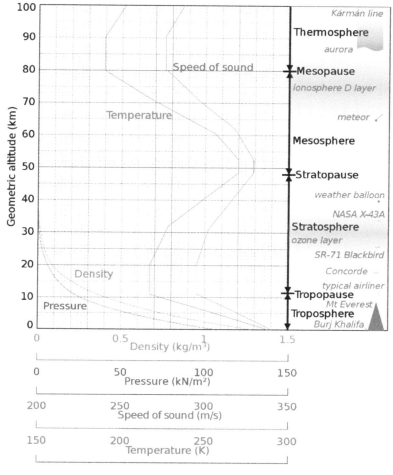

Figure A.1. Temperature, speed of sound, density, and pressure profiles of the atmosphere. Credit: Wikipedia: Cmglee (CC BY-SA 3.0).

greenhouse effect. In general, the state of the atmosphere is defined by the radiative balance between the Earth, its atmosphere, and the Sun. Given an incident power from the Sun, the phenomena that must be considered are:

- Emission, which is due to molecular transitions caused by the temperature of the atmosphere.
- Absorption of the incident radiation, which is also due to molecular transitions.
- Diffusion, which is the redistribution of energy in all directions.

For astrophysical applications we are interested in transmission properties, but also in stability, and the emission and refractive properties. We must consider the emission of the atoms and molecules involved both as continuous emission (blackbody emission) and as emission or absorption lines. Normally the atmosphere is divided into parallel layers and the radiative transport between one layer and the other is considered then to determine global quantities such as emission and transmission. Let's calculate the power that affects the outer layer of the atmosphere (coming from the Sun) and then we will see how it is transmitted inside. The power emitted by the Sun is ($T_{sun} = 5780$ K; $r_s = 6.96 \times 10^8$ m):

$$W = \sigma T_{sun}^4 \cdot 4\pi r_s^2 = 3.86 \times 10^{26} \text{ W.} \qquad (A.1)$$

The solid angle with which the Sun sees the Earth is ($r_T = 6.37 \times 10^6$ m; Earth–Sun distance $d = 1.5 \times 10^{11}$ m):

$$\Omega_T = \frac{\pi r_T^2}{d^2} = 5.67 \times 10^{-9} \text{ sr.} \qquad (A.2)$$

The projected area of the Earth is:

$$A_T = \pi r_T^2 = 1.28 \times 10^{14} \text{ m}^2. \qquad (A.3)$$

Hence the solar constant CS is:

$$\text{CS} = \frac{W}{4\pi} \cdot \Omega_T \cdot \frac{1}{A_T} \sim 1365 \text{ Wm}^{-2} \rightarrow \frac{CS}{4} \sim 342 \text{ Wm}^{-2}. \qquad (A.4)$$

CS represents the average incident flow on the upper layer of the atmosphere on the projected Earth.

A.2 Radiative Transfer

A.2.1 Absorbance

Consider a medium. Take a portion of thickness ds and consider radiation I at a wavelength λ incident on it. We can define $a(\lambda)$, the absorption coefficient (with dimensions: $[a(\lambda)] = 1/m$), as the fraction of radiation I that is absorbed along an infinitesimal path. We have:

$$I(\lambda, s) + dI(\lambda, s) = I(\lambda, s) - I(\lambda, s)a(\lambda)ds. \qquad (A.5)$$

Thus:

$$\frac{dI(\lambda, s)}{ds} = -a(\lambda) \cdot I(\lambda, s) \qquad (A.6)$$

and so:

$$I(\lambda, s) = I(\lambda, 0) \cdot e^{-sa(\lambda)}. \qquad (A.7)$$

The absorption coefficient $a(\lambda)$ contains the physics of the constituents of the atmosphere. Let's define the opacity τ and its differential $d\tau$ as:

$$d\tau(\lambda) = a(\lambda) \cdot ds. \qquad (A.8)$$

Equation (A.7) becomes:

$$\frac{dI(\lambda, s)}{d\tau(\lambda)} = -I(\lambda, \tau) \rightarrow I(\lambda, \tau) = I(\lambda, 0) \cdot e^{-\tau(\lambda)}. \tag{A.9}$$

If we consider p different species that contribute to the absorption, we have to sum up their contributions:

$$\tau(\lambda) = \sum_{i=0}^{p} \tau_i(\lambda). \tag{A.10}$$

A.2.2 Emission

Absorbance A is defined as the fraction of radiation absorbed in a given ds path. Absorbance and the absorption coefficient for a homogeneous medium are linked by:

$$A_{ds}(\lambda) = 1 - e^{-a(\lambda)ds} = 1 - e^{-\tau(\lambda)}. \tag{A.11}$$

For a blackbody (BB), A = 1 for every λ. However, a blackbody has an emissivity of $\varepsilon = 1$. Indeed, we have Kirchhoff's law, which states that:

$$A(\lambda) = \varepsilon(\lambda). \tag{A.12}$$

So the emission coming from the ds element will be ($BB(\lambda, T)$ is the blackbody emission brightness):

$$dI^{\mathrm{emiss}}(\lambda, s) = A_{dS}(\lambda) \cdot BB(\lambda, T) = (1 - e^{-a(\lambda)dS}) \cdot BB(\lambda, T) \tag{A.13}$$

and so:

$$I(\lambda, \tau) = (1 - e^{-\tau(\lambda)}) \cdot BB(\lambda, T) + I(\lambda, 0) \cdot e^{-\tau(\lambda)}. \tag{A.14}$$

If we now divide the atmosphere into multiple layers i of opacity τ_i and temperature T_i, each of which absorbs the emission of the previous layer $i - 1$ and emits towards the next layer $i + 1$, we can write the radiation through the ith layer using the simplified equation of radiative transfer:

$$I_i(\lambda) = (1 - e^{-\tau_i(\lambda)}) \cdot BB(\lambda, T_i) + I_{i-1}(\lambda) \cdot e^{-\tau_i(\lambda)}. \tag{A.15}$$

Once the information on T, P, density, and molecular abundances in the various layers has been entered, the solution is found numerically. It starts with:

$$I_0(\lambda) = BB(\lambda, T_0) \tag{A.16}$$

and it continues until it has cycled through all of the layers.

Once the solution has been found, it is possible, for example, to calculate the transmissivity from the opacity:

$$t(\lambda) = e^{-\tau(\lambda)} \tag{A.17}$$

or the brightness temperature that derives from Planck's law (Figure A.2):

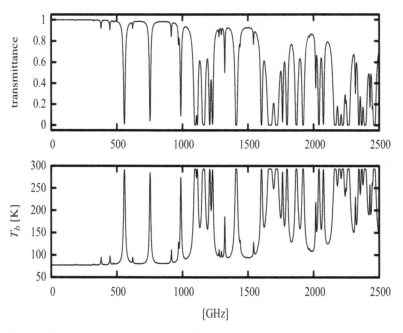

Figure A.2. Transmission and emission of the atmosphere as a function of frequency.

$$T_b = \frac{hc}{k_B \lambda} \left[\ln\left(1 + \frac{2hc^2}{I(\lambda)\lambda^5} \right) \right]^{-1} . \tag{A.18}$$

A.3 Atmospheric Diffusion

The scattering of radiation by the constituents of the atmosphere also depends on the relative sizes of the particles and the wavelength:

- Gas molecules have dimensions $d \sim 0.1$ nm.
- Aerosol particles have dimensions $d \sim 10$ nm $\rightarrow 10$ μm.
- Water droplets have dimensions $d \sim 10$ μm $\rightarrow 100$ μm.

For this reason, different bands of the electromagnetic spectrum behave differently when in contact with the atmosphere:

- For $d \ll \lambda$, there is Rayleigh molecular scattering.
- For $d \gg \lambda$, there are scatterings related to geometric optics.
- For $d \sim \lambda$, a detailed account of the scattering must be made using detailed models.

The radiative transfer equation must include the different atomic or molecular transitions (and their enlargement). These can be included in the determination of how the optical thickness of the atmosphere varies as a function of the scattering solid angle, frequency by frequency.

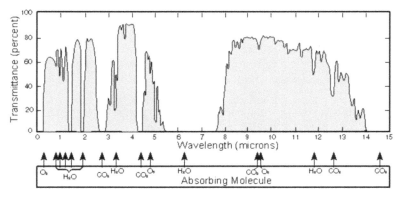

Figure A.3. Atmosphere transmission spectrum with the molecules responsible for the absorption labeled. Credit: Wikipedia: (Public Domain) https://commons.wikimedia.org/w/index.php?curid=34818020.

- At high frequencies (in the UV, X-ray, and γ-ray), the ionized medium of the ionosphere/stratosphere blocks all photons. In the UV, absorption is mainly linked to the presence of O_2, O_3, and N_2 and their electronic transitions.
- Electronic transitions dominate in the visible. In the visible and near-infrared, we have H_2O, CO_2, and O_3 and their electronic and vibrational transitions.
- Vibrational molecular transitions are important in the infrared. In the infrared, we have vibrational transitions of H_2O and CO_2 and rotational transitions of O_3.
- Rotational molecular transitions are important in the far-infrared and microwave frequencies.

In the microwave regime we have rotational transitions of O_2 and rotational transitions of H_2O. The lines widen due to the motion of the molecules (Doppler broadening), the collisions (pressure broadening), the long tails of the lines themselves, and changes in pressure and temperature that also cause continuous absorption (Figure A.3).

In almost all software used to calculate the atmospheric transmission, the amount of H_2O must be specified, as it is highly variable. In the infrared and microwave regime, the H_2O transitions that dominate are the vibrational/rotational ones, while in the UV they are the electronic ones. H_2O is concentrated in the first layer of the atmosphere (the troposphere) and is the main adversary of almost all fields of astronomy. It is therefore necessary to build observatories at high altitude and in dry places. The Atacama desert and Antarctica are among the best sites in the world.

An alternative is to use satellite or long-duration balloons, which are stratospheric balloons that can perform flights of ~15 days.

A.4 Variability and Fluctuations

Another problem caused by the atmosphere is its variability, which is mainly linked to random changes in temperature (and density) generated by turbulence on large angular scales (see Kolmogorov's theory). In microwave frequencies, this adds low-frequency noise because the emission of the atmosphere varies with time. In the visible, this generates temporal variations of the refractive index of the atmosphere. This produces variations in the position of the sources on the sky and in fact reduces the angular resolution. We can quantify this effect through the Fried parameter r_0, which estimates the coherence of a light wave. The Fried parameter depends on the wavelength λ, on the zenith angle z, and on the refractive index of the atmosphere (which in turn depend on pressure and temperature) and it is measured based on the effect of the atmosphere on radiation or with probes mounted on balloons. In the visible, $r_0 \sim 10$ cm

$$r_0 = 0.185 \cdot \lambda^{\frac{6}{5}} \cdot \cos(z)^{3/5} \left(\int C_n^2(h)dh \right)^{-3/5}. \tag{A.19}$$

An undisturbed wave front is disturbed by vortices present in the atmosphere. "Seeing" generates three effects (which are studied empirically):

- Enlargement: angular resolution is no longer dominated by diffraction, but by r_0,

$$1.22\frac{\lambda}{D} \rightarrow 1.22\frac{\lambda}{r_0}. \tag{A.20}$$

- The source moves with a variance that is also determined by r_0 and by the diameter D:

$$\sigma_x^2 = \sigma_y^2 = 0.18 \cdot D^{1/3}r_0^{-5/3}. \tag{A.21}$$

- Scintillation: the source changes intensity. The relative variance of its brightness will depend on the wavelength λ and on the zenith angle z:

$$\frac{\sigma_I^2}{I^2} = 19.12 \cdot \lambda^{-\frac{7}{6}} \cos(z)^{-11/6}. \tag{A.22}$$

AAS | IOP Astronomy

Experimental Astrophysics

Elia Stefano Battistelli

Appendix B

Astronomical Coordinates Recap

B.1 Horizontal Coordinates System

In order to determine the position of an object on the celestial sphere, we need two coordinates. The horizontal coordinate system (see Figure B.1) sets these two coordinates and uses the horizon as a fundamental plane. For this reason, the coordinates depend on the position of the observer. The two coordinates are named azimuth and altitude (or elevation). For a northern observer, the azimuth can be measured in degrees and ranges between 0° and 360°, from north (0°), to east (90°), to south (180°), to west (270°). The altitude (or elevation) measures the angular distance from the horizon. It ranges from −90° and +90°: 0° elevation is the position of the horizon, while +90° is the position of the *zenith* (above the observer's head) and −90° is the *nadir* (below the observer's feet). We thus have:

- Azimuth: 0°/360° from north to east
- Altitude (elevation): 0°/(±)90° measured from the horizon.

The zenith angle (the angular distance from the zenith) is the complementary angle, at 90°, of the elevation. A celestial object transits (or culminates) when it crosses its meridian. The meridian is the celestial circle crossing both of the celestial poles (which are extensions of the Earth's poles into the sky) and the zenith and nadir.

Clearly, the horizontal system cannot uniquely identify the coordinates of astronomical objects. However, it is a very useful system, in practice, for the movement of telescopes (large telescopes in particular).

If a celestial object is observed from Earth's surface, it is observed through a layer of atmosphere. The thickness of the atmosphere, the airmass (*am*), depends on the airmass at the zenith (am_z, assumed to be 1) and the zenith angle z according to the so-called *secant law*:

$$am = \frac{am_z}{\cos(z)} = \sec(z).$$
(B.1)

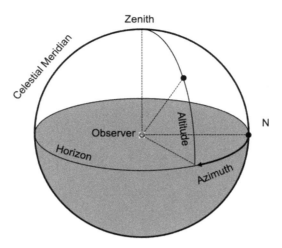

Figure B.1. The horizontal coordinates system.

The horizontal system allows for the effect of the atmosphere on an observation to be monitored as a function of the elevation. Observations of object that appear close the horizon are clearly worse because the radiation passes through a thicker atmosphere.

B.2 Equatorial Coordinates System

The equatorial coordinate system is a celestial coordinate system widely used in astronomy (Figure B.2). It is based on a projection of the Earth's coordinates onto the celestial sphere where the stars are held fixed. The two coordinates are obtained by projecting the Earth's axis onto the celestial sphere, delineating a north celestial pole and a south celestial pole, as well as the celestial equator. The declination (δ or DEC) ranges from $-90°$ and $+90°$ ($+90°$ being the north celestial pole, NCP, and $-90°$ being the south celestial pole, SCP). The right ascension (α or RA) is measured in hours, minutes, and seconds and ranges between 0 and 24 h. RA = 0 h is the position of the Sun on the spring equinox, the point at which the equatorial plane crosses the ecliptic:

- Dec: $-90° < \delta < +90°$ ($+90° =$ NCP; $-90° =$ SCP). It is measured in degrees, minutes, seconds ($°$, $'$, $''$).
- RA: $0\,\text{h} < \alpha < 24\,\text{h}$. It is measured in hours, minutes, seconds (h, m, s).

The sidereal time is the time dictated by the Earth's rotation with respect to the fixed stars. Due to the revolution of the Earth around the Sun, the sidereal day is 23 h 56 m 4.09 s and it is widely used by astronomers to locate celestial objects.

The equatorial system is a common system for all observers. However, due to axis precession, one has to specify the year of the coordinates. Unless it is specified differently, we will consider coordinates at the year 2000 (J2000).

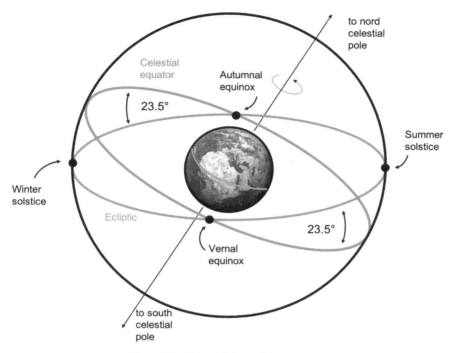

Figure B.2. Equatorial coordinate system.

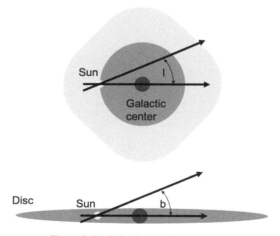

Figure B.3. Galactic coordinate system.

B.3 Galactic Coordinates System

Our Galaxy, the Milky Way, is a barred spiral galaxy. The Galactic coordinate system (see Figure B.3) has been defined starting from the consideration that the solar system is at the periphery of our Galaxy and that the Milky Way can by observed as a plane (the Galactic plane) in the celestial sphere. These angular coordinates describe the position of an object in the the Milky Way in terms of a Galactic latitude, b, the angular distance perpendicular to the Galactic plane toward the north (north Galactic pole, $+90°$) or toward the south (south Galactic pole, $-90°$). Equivalently, the Galactic longitude, l, is the angular distance of an object along the Galactic equator from the Galactic center, as measured from the Sun.

- Galactic longitude: $0° < l < 360°$
- Galactic latitude: $-90° < b < +90°$.

The Galactic center and Galactic anticenter have the coordinates $(0°, 0°)$ = Galactic center and $(180°, 0°)$ = Galactic anticenter. The Galactic coordinate system is important for extragalactic observations in which we want to estimate the angular "distance" of an object from the center of the Galaxy, and we want to consider the astrophysical contamination of the Galaxy to our observations.

Experimental Astrophysics

Elia Stefano Battistelli

Appendix C

Laboratory Activities

In the following, there is a list of laboratory activities one could undertake:

C.1 Test of the Nyquist theorem through the use of a signal generator and a digital oscilloscope.

C.2 Test of the flux $1/r^2$ law using an LED and a photodiode.

C.3 Signal extraction from noise (and test of the flux $1/r^2$ law) using an LED, a photodiode, and a lock-in amplifier.

C.4 Measurement of the Boltzmann constant from the measurement of the Johnson noise of an ambient-temperature resistor.

C.5 Measurement of the speed of light using a pulsed LED, a mirror, and an oscilloscope.

C.6 Construction of a negative-feedback proportional, integral, differential (PID) loop using a microcontroller such as an Arduino microcontroller.

C.7 Preparation of an observational proposal to an existing astronomical facility.

Roof activities:

C.8 Measurement of the sky background temperature at radio wavelengths with a TV parabolic antenna, measuring the secant law of the atmospheric emission as a function of the elevation. Is it consistent with 2.7 K?

C.9 Use of an optical telescope: preliminary calibration observations (bias, dark, flat), photometry, and spectrometry of bright astronomical sources.

C.1 Test of the Nyquist Theorem through the Use of a Signal Generator and a Digital Oscilloscope

Phase 0: Introduction

This experiment makes use of an oscilloscope. An oscilloscope allows us to measure an electrical signal as a function of time by generating an X–Y graph in two dimensions. An oscilloscope performs repeated scans by measuring a voltage Y versus time t (X). The sizes of the axes are adjustable and are measured in volts per

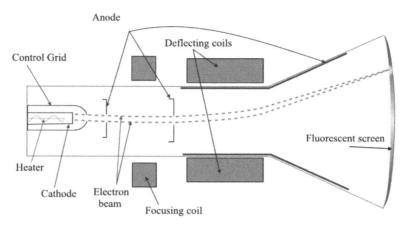

Figure C.1. Cathode-ray tube, the basis for the design of an analog oscilloscope. Credit: Wikipedia: Theresa Knott (CC BY-SA 3.0).

division (Y) or s per division (X). Thanks to the ability to vary the size of the time axis (for example, as a multiple of the signal period), an oscilloscope can measure periodic signals. And, the use of a trigger (which starts the scan depending on an event) allows the oscilloscope to detect impulsive signals as well as periodic ones. The signals are inserted into the BNC connectors (or for high frequencies, the SMA or other connectors). Some oscilloscopes have multiple input channels with the ability to graph one against another. An analog oscilloscope is based on an electronic cannon, a cathode tube under vacuum in which electrons travel, with a phosphorus-coated screen that lights up when hit by an electron flow and horizontal and vertical deflection plates regulated by the input signal (Figure C.1).

A digital oscilloscope, however, is based on the use of analog-to-digital conversion (ADC): it needs random access memory (RAM), a microprocessor that allows us to analyze data in real time (or almost in real time; Figure C.2). Oscilloscopes have the ability to perform an auto-set, by which the oscilloscope "decides" the settings based on the signal.

Phase 1: Preliminaries

- Prepare yourself with a laboratory notebook to take notes in.
- Turn on the devices (unless you need to turn them off, leave them on for the whole experiment).
- Connect the tools.
- Design the experiment and, if useful, write out a list of the steps.

Phase 2: Using the Tools

- Generate sine, square, and triangular waves and display them on the oscilloscope by varying the parameters of the oscilloscope. Try different frequencies: 10 Hz, 1 kHz, 100 kHz, and 10 MHz.

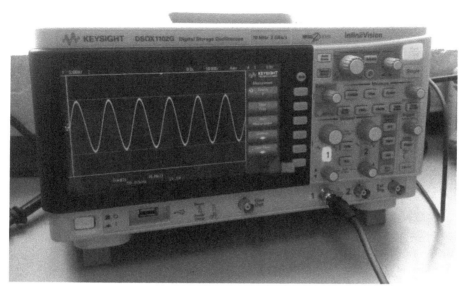

Figure C.2. A Keysight digital[3] oscilloscope.

- Acquire the waveform and learn how to save waveform data to a file. Import the file into data processing software (e.g., Origin[1] or Excel)[2].
- Check the data acquisition Δt. This is important for the sampling theorem.

Phase 3: Quantizations

- Acquire waveforms of the following types:
  - Sinusoidal
  - Square
  - Triangular
  - Noise.
- For each of them, build a histogram of the voltages and describe its form (Figure C.3).

Phase 4A: Quantization

- Decrease the amplitude of the signal without increasing the vertical amplification of the oscilloscope (leaving the bin number of the histogram unchanged).
- In this phase, the effects of the quantization of the sampled data will be visualized.

---

[1] https://www.originlab.com/
[2] https://www.microsoft.com/
[3] https://www.keysight.com/

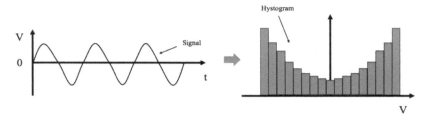

Figure C.3. Typical sinusoidal signal acquired by an oscilloscope and its histogram.

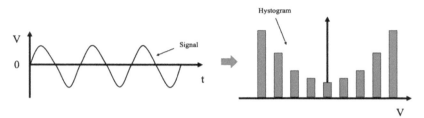

Figure C.4. Sinusoidal signal not well measured, with quantization problems.

- Evaluate from the data the quantization step of the ADC of the oscilloscope, compare it with the expected theoretical value, and check if it depends on the vertical amplification.

Phase 4B: Quantization
- Increase the vertical amplification of the oscilloscope, leaving the amplitude of the signal unchanged (always leaving the bin number of the histogram unchanged).
- Also in this phase, the effects of the quantization of the sampled data will be visualized.
- As above, evaluate from the data the quantization step of the ADC of the oscilloscope, compare it with the expected theoretical value, and check if it depends on the vertical amplification (Figure C.4).

Phase 5: Sampling
- Reset the waveform generator and oscilloscope to conditions in which there is no data loss.
- Without changing the horizontal scale factor of the oscilloscope, adjust the waveform generator to increase the signal frequency.
- At each frequency, take note of the "true" frequency of the signal and then analyze the data to determine what frequency is measured by the data.

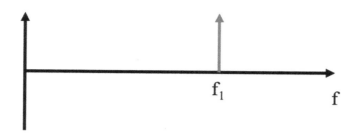

Figure C.5. Input signal frequency f_1.

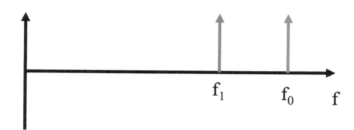

Figure C.6. Sampling frequency f_0.

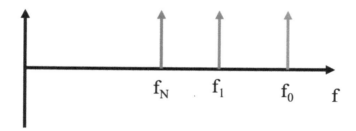

Figure C.7. Nyquist frequency f_N.

Phase 6: Sampling Calculation

- Determine the theoretical frequencies involved: calculate the Nyquist frequency and half of the Nyquist frequency.

Phase 7: Proof the Sampling Theorem

- We apply what we learned about the Nyquist–Shannon theorem with single-frequency signals:
  - We have a signal at a frequency f_1 (Figure C.5).
  - We sample it with a sampling frequency f_0 (Figure C.6).
  - We then have a Nyquist frequency $f_N = f_0/2$ (Figure C.7).

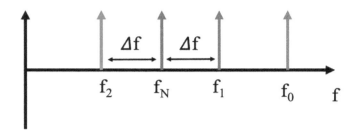

Figure C.8. The distance between f_2 and f_1 with respect to f_N is the same Δf.

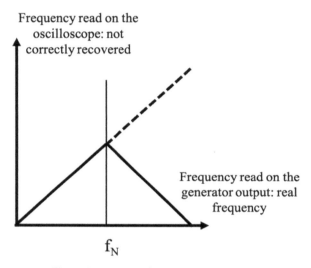

Figure C.9. Nyquist frequency determination.

○ If the frequency of the signal is $f_1 > f_N$, we then have a *mirroring* of f_1 into f_2, Δf away from f_N. The recovered signal is thus grossly wrong: it is indistinguishable from a real signal at frequency $f_2 = f_N - \Delta f$ (Figure C.8).

Phase 8: Proof of the Sampling Theorem: a Demonstration

• Measure the Nyquist frequency and compare it with respect to the theoretical value through the construction of the following graph. Increasing the input frequency (the x-axis), and calculate the recovered frequency (the y-axis). This allows for the determination of the slope of the straight lines before and after the Nyquist frequency, and the determination of the Nyquist frequency itself. All of the errors must be propagated when determining the Nyquist frequency (Figure C.9).

C.2 Test of the Flux $1/r^2$ Law Using an LED and a Photodiode

Phase 0: Preliminaries

For this experiment we will use the following instrumentation:
- An LED light emitter
- A photodiode detector
- A square-wave generator (for the LED)
- An oscilloscope
- A rail to vary the distance between the LED and photodiode
- A PC with an acquisition system
- Accessories:
  - A multimeter
  - BNC and "T" connectors
  - A voltage generator (power supply) for the photodiode
  - An Allen key for the rail.

Aims of the experiment:
- Verification of flux law $F \propto 1/r^2$
- Estimation and verification of brightness conservation
- Responsivity measures
- Analysis of noise and other contributions
- Measurement of the time constant (optional).

Most LED luminosities are published on their datasheet and are expressed in candelas. The candela (cd) is an estimate of the light power emitted per unit of solid angle from a source at the maximum response of a human eye (not dark adapted). Another useful quantity is the lumen (lm), which is a measurement of total intensity. An isotropic source, for example of 1000 lm, has a radiant intensity in cd as in the following:

$$I = \frac{1000 \text{ lm}}{4\pi} \sim 80 \text{ cd}. \tag{C.1}$$

The candela is a unit of measure of radiant intensity equivalent to the power of 1/683 W sr^{-1} from a monochromatic source at 555 nm. If the emission is at a different wavelength, we have to convert it according to the brightness function that describes the relative efficiency of the response of the human eye. If the response is broadband, we have to perform an integration. A source that, in SI units, emits $I_e(\lambda)$ [W sr^{-1}] will have a radiant intensity $I_v(\lambda)$ [cd] of:

$$I_v(\lambda) = 683 \cdot \bar{y}(\lambda) \cdot I_e(\lambda) \rightarrow I_\nu = \int 683 \cdot \bar{y}(\lambda) \cdot I_e(\lambda) d\lambda \approx \eta \cdot I_e. \tag{C.2}$$

The values of η are typically found in the datasheet that accompanies an LED.

Phase 1: Understanding a Datasheet

It is fundamental to learn how to read a device datasheet. For an LED, for instance, in addition to the value of η, we need the following information:
- The absolute maximum ratings (e.g., a DC forward current of 7 mA)
- The physical dimensions
- The type of power supply and its typical values (e.g., a DC forward current of 2 mA)
- The response characteristic (luminous intensity versus current)
- The response time.

For the photodiode, we need the following information:
- The maximum ratings (e.g., a maximum DC supply of +/− 18 V)
- The type of power supply
- Linearity information
- Efficiency information
- Transfer curves
- Noise information.

Phase 2: Choosing Parameters

The parameters listed on the datasheet must be correlated with those of the power-supply circuits (e.g., by measuring R with the open-circuit multimeter).

The most important parameter is the current to be passed to the LED, considering the power-supply circuit of the LED. For example, if we want to run a current of 4 mA in the LED, we have to consider the I–V curve of the LED on the datasheet (Figures C.10 and C.11).

And thus, we must derive the voltage at its ends and the generator function to be applied:

$$I_F = 4\,\text{mA} \rightarrow V_F = 1.8\,\text{V} \rightarrow V_{GEN} = V_F + I \cdot R = 3.67\,\text{V}. \qquad (C.3)$$

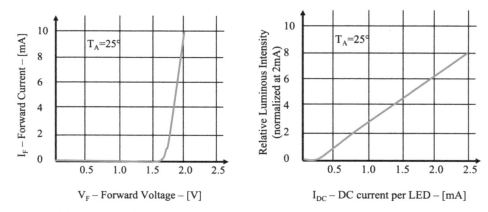

Figure C.10. Left: I–V of the LED. Right: radiated power vs. current for the LED.

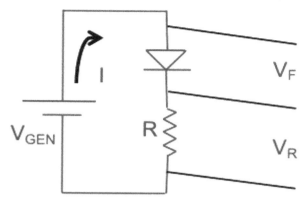

Figure C.11. Power circuit of the LED.

Once the LED is powered, it will be necessary to verify, for example, that $V_F \sim 1.8$ V. Given the I_F current, we will be able to estimate the radiant intensity

$$I_F = 4 \text{ mA} \rightarrow I_V = 3 \cdot 2.3 \text{ mcd} \tag{C.4}$$

and thus

$$I_e = \frac{I_V}{\eta} = 47.6 \, \mu\text{Wsr}^{-1}. \tag{C.5}$$

Phase 3: Using the Instruments

- Prepare yourself with a laboratory notebook to take notes in.
- Turn on the devices (unless you need to turn them off, leave them on for the whole experiment).
- Connect the tools.
- Design the experiment and, if useful, write out a list of the steps.
- Generate sine, square, and triangular waves and display them on the oscilloscope by varying the parameters of the oscilloscope. Try different frequencies: 10 Hz, 1 kHz, 100 kHz, and 10 MHz.
- Acquire the waveform and learn how to save waveform data to a file. Import the file into data processing software.

Phase 4: Connecting the Photodiode

- Connect the BNC output of the photodiode to one of the oscilloscope inputs.
- Supply the photodiode with a voltage:
  - +9 V (red-low)
  - 0 V/COM, connection to the center
  - −9 V connection at the top
- Adjust the amplitude and sampling of the oscilloscope in order to have $f_N \gg$ 1 kHz.

- Turn on the photodiode and observe the measured signal. What signal is it? What's the frequency of the maximum noise signal? Roughly evaluate its amplitude.
- Connect the PC and prepare the acquisition system.
- Look again at the detected signal and verify that it corresponds with what is displayed in the oscilloscope.
- Take a measurement of dark signal and compare it with the previously acquired signal, possibly in frequency space.

Phase 5: Connecting the LED

- Connect the LED to the main output of the waveform generator according to the amplitude values previously determined (add a DC offset to make the negative wave front null!).
- Connect the generator output to the other input of the oscilloscope using the "T" connector.
- How should the oscilloscope be set? In DC or AC?
- Verify that the photodiode sees a signal in phase with the generator output.
- Try to change the power-supply frequency of the LED and select one that is not a multiple of the frequencies at which you observed noise previously (a good frequency should be 280 Hz ... check!).
- Move the detector closer until it is about 10 cm from the LED and observe the measured signal (peak-to-peak). Roughly evaluate the signal-to-noise ratio S/N.
- Move the detector away from the LED to the point at which S/N < 1 (or to the end of the rail).

Phase 6: Measuring $F(r)$

- Measure the signal from the LED for different distances between the photodiode and the LED, within the previously determined range of distances.
- For each distance, evaluate the peak-to-peak amplitude of the signal and the S/N.
- Evaluate the flux law as a function of the distance $F = F(r)$ by reporting the values in a double-logarithmic graph.
- Note: an important systematic effect of this measure is the orientation of the LED and the photodiode. Measurements performed "in axis" are correct, but if there is a misalignment between the two, the measurements will be incorrect. Determine when the effects of misalignment become important.

Phase 7: Measuring R

- For each of the measurements made previously, evaluate the solid angle with which the LED "sees" the photodiode using the dimensions of the photodiode and the distance between the source and the receiver.

- Determine the responsiveness in V/W by evaluating the signal from peak to peak, at least over three periods, and propagating the uncertainties.
- Estimate the brightness based on the measured signal and the responsiveness by deriving the solid angle with which the photodiode sees the LED.

Phase 8: Measuring Noise

- Carry out the measurement both in dark conditions (covering the photodiode in order to avoid disturbing signals from the lamps) and with the photodiode open to light, but in the absence of signal.
- Measure the noise by appropriately amplifying the voltages detected by the oscilloscope.
- Acquire the signal due to noise and determine the noise-equivalent voltage (NEV) from the standard deviation of the "time stream."
- Derive the noise-equivalent power (NEP)[4] from the NEV and responsiveness:

$$NEP = \frac{NEV}{R}. \tag{C.6}$$

- OPTIONAL: Derive the noise level on the Fourier transform of the signal. This can be done either by checking the appropriate constants entered by the software that makes the Fourier transform (e.g., Origin)[5], or by determining the responsiveness and noise on the Fourier transform of both the peak of the modulated signal and the noise baseline in order to simplify any multiplicative factors.

Phase 9: Measuring the Time Constant (Optional)

- Put the system in a configuration in which the S/N is optimal.
- Vary the modulation frequency of the LED signal.
- Observe the signals measured by the photodiode on the oscilloscope.
- By increasing the frequency, it will be clear at a certain point that the rising and falling wave fronts are not instantaneous.
- Acquire the signals for some frequencies and analyze them with functions fit with different time constants (that will be compared) for the ascent and descent,

$$V(t) = A[1 - e^{-t/\tau}] + B. \tag{C.7}$$

Note: on a semi-logarithmic scale, $\log(V(t)-B)$ versus t is a straight line, from the slope of which it is possible to obtain τ.

---

[4] To be rigorous, the NEP is a spectral quantity measured in W $Hz^{-1/2}$. Here, however, we will consider the integrated function of it.
[5] https://www.originlab.com/

C.3 Signal Extraction from Noise (and Test of the Flux $1/r^2$ Law) Using an LED, Photodiode, and a Lock-in Amplifier

Phase 0: Preliminaries

For this experiment we will use:
- An LED light emitter
- A photodiode detector
- A square-wave generator for the LED
- An oscilloscope
- A rail to vary the distance between the LED and photodiode
- A PC with acquisition system
- Accessories:
  - A multimeter
  - BNC and "T" connectors
  - A voltage generator (power supply) for the photodiode
  - An Allen key for the rail
- Lock-in amplifier.

Aims of the experiment:
- Use of the synchronous demodulation method for the extraction of the noise signal
- Verification of the law of flow
- Responsivity measures
- Analysis of noise and different contributions.

Phase 1: Double-check the Previous Experiment (C.2)

- Write down the values of responsiveness and their uncertainties.
- Write down the S/N values for the different LED versus photodiode positions.
- Make a note of the measured noise, NEV, and NEP values.
- Evaluate the different sources of noise and noise in the signals acquired during the previous exercise.

Phase 2: Turning On the Lock-in Power Supply

- Turn on the power supply and adjust it to produce two symmetrical voltages with respect to zero, equal to +9 V and −9 V. If necessary, check with the multimeter.
- Turn off the power supply.
- Connect the power cable to the lock-in amplifier and connect the three banana connectors as follows:
  - the red banana connector at the positive output of the power supply (red plug)

Figure C.12. Lock-in amplifier.

○ the black banana connector at the negative output of the power supply (blue plug)
○ the yellow banana connector at the mass output (COM) of the power supply (black plug)
○ Turn on the power supply.

Phase 3: Connecting the Lock-in Amplifier

- Connect a sinusoidal signal from the generator to the lock-in amplifier input (red BNC) and to the oscilloscope via a "T" connector (Figure C.12).
- Connect the TTL output of the generator to the reference input of the lock-in amplifier (blue BNC).
- Connect the output from the internal white BNC (the multiplier output) to the oscilloscope.
- Evaluate the shape of the signal on the oscilloscope: what shape does it have?

Phase 4: Calibrating the Lock-in Amplifier

- Connect the lock-in filter output (the external white BNC) to the oscilloscope. What shape does the signal have?
- Vary the amplitude of the input signal from 50 mV peak-to-peak to 500 mV peak-to-peak.
- Measure the output signal for each incoming signal using the oscilloscope's "average" function.
- Record the output signal data as a function of the input signal in a table and in a graph.
- Check the linearity and estimate the proportionality constant C between the input and output signals.

Phase 5: Connecting the Lock-in Amplifier to the Photodiode

- Connect the photodiode signal to the lock-in amplifier input signal (red BNC).

- Connect the TTL output of the generator to both the LED and the reference input of the lock-in amplifier (blue BNC) using a "T" connector.
- Observe the signal exiting the lock-in amplifier and evaluate its S/N.

Phase 6: Determining the Flux Dependence on Distance

- Measure the amplitude of the photodiode signal by varying its distance to the LED using the entire rail.
- For each position, evaluate the measured signal and the S/N. Also evaluate the error related to the stability of the lock-in amplifier output.
- Record the signal data as a function of distance in a table and a graph and confirm the trend $F = F(r)$ by evaluating the flow law as a function of the distance (create a double-logarithmic plot).

Phase 7: Measuring the Responsivity

- For each of the measurements made previously, evaluate the solid angle with which the LED "sees" the photodiode based on the dimensions of the photodiode and the distance between the source and the receiver.
- Determine the responsiveness in V/W by evaluating the signal from the average signal coming out of the lock-in amplifier.
- Consider the lock-in amplifier calibration constant and compare the responsivity values obtained both with and without the presence of the lock-in amplifier.

Phase 8: Measuring the Noise

- Carry out the measurement both in dark conditions (covering the photodiode in order to avoid disturbing signals from the lamps) and with the photodiode open to light but in the absence of signal.
- Measure the noise by appropriately amplifying the voltages detected by the oscilloscope.
- Acquire the noise signal and determine the noise-equivalent voltage (NEV) from the standard deviation of the "time stream" output at the lock-in amplifier.
- Derive the noise-equivalent power (NEP) from the NEV and responsiveness:

$$\text{NEP} = \frac{\text{NEV}}{R}. \tag{C.8}$$

- Consider the lock-in amplifier calibration constant and compare the NEP values obtained both with and without the presence of the lock-in amplifier.

C.4 Measurement of the Boltzmann Constant from the Measurement of the Johnson Noise of an Ambient-temperature Resistor

Phase 0: Preliminaries

Johnson noise arises from fluctuations electrical charges in a conductor and depends on the temperature and electrical resistance of the conductor. The power spectrum of current fluctuations is white up to a frequency (band) $= 1/\tau$ (the band is decided by any capacities or inductances in the circuit). The total power (current) in a limited band is therefore:

$$\langle \delta I^2 \rangle = \int_{\Delta f} S(f)df = 4k_{\mathrm{B}}TG\Delta f \tag{C.9}$$

where k_{B} is the Boltzmann constant, T is the temperature, and G is the conductance, the inverse of the resistance R. In units of voltage, we have:

$$\langle \delta V^2 \rangle = R^2 \langle \delta I^2 \rangle = 4k_{\mathrm{B}}TR\Delta f. \tag{C.10}$$

The power, on the other hand, is independent of the resistance and sets a minimum value for a one-T circuit (e.g., at 300 K, $W = 4 \times 10^{-21}$ W Hz$^{-1/2}$):

$$P = \sqrt{\langle \delta V^2 \rangle \langle \delta I^2 \rangle} = \sqrt{\frac{R}{R}} 4k_{\mathrm{B}}T\Delta f \approx k_{\mathrm{B}}T\Delta f. \tag{C.11}$$

These are small values (10^{-24} W Hz$^{-1/2}$ at 0.1 K) but they are useful for precision calibrations (e.g., bolometers). The voltage power spectrum is, therefore, for small-enough frequencies:

$$S_{\mathrm{V}}(f) = \frac{4k_{\mathrm{B}}TR}{1 + (2\pi f\tau)^2} = 4k_{\mathrm{B}}TR. \tag{C.12}$$

From Parseval's identity we have:

$$\int_{\Delta f} S_{\mathrm{V}}(f)df = \langle \delta V^2 \rangle = 4k_{\mathrm{B}}TR\Delta f. \tag{C.13}$$

So, if we measure the resistance of a resistor (at its ends) with an oscilloscope we get an rms:

$$\sqrt{\langle \delta V^2 \rangle} = \sqrt{4k_{\mathrm{B}}TR\Delta f}. \tag{C.14}$$

We want to use this noise in order to infer the value of the Boltzmann constant k_{B}. If:

- $R = 10$ MΩ
- $T = 300$ K
- $\Delta f = 1$ kHz

we get:

$$\sqrt{\langle \delta V^2 \rangle} = \sqrt{4k_B TR\Delta f} = 12.8\ \mu V. \tag{C.15}$$

Thus, we expect to measure fluctuations with a standard deviation of $12.8\ \mu V$ at the ends of a $10\ M\Omega$ resistor at ambient temperature. This value is clearly too small for the oscilloscope to detect. It is necessary to amplify the signal with a high gain (e.g., $A = 1000$) over the whole band.

$$V_{out}(f) = A(f) \cdot V_{in}(f) \tag{C.16}$$

So:

$$S_{V_{out}}(f) = A^2(f) \cdot S_{V_{in}}(f) = 4k_B TR \cdot A^2(f) \tag{C.17}$$

and so:

$$\langle \delta V_{out}^2 \rangle = \int_{\Delta f} S_{V_{out}}(f) = 4k_B TR \cdot \int_0^\infty A^2(f) df. \tag{C.18}$$

If, for example:

$$A = \begin{cases} 1000 & \text{for} \quad 0 < f < 1\ \text{kHz} \\ 0 & \text{for} \quad f > 1\ \text{kHz} \end{cases} \tag{C.19}$$

we get:

$$\langle \delta V_{out}^2 \rangle = 4k_B TR \cdot A^2(f) \cdot \Delta f \tag{C.20}$$

and so

$$\sqrt{\langle \delta V_{out}^2 \rangle} = 1000 \cdot \sqrt{4 \cdot k_B \cdot 300\ K \cdot 10\ M\Omega \cdot 1\ kHz} = 12.8\ mV. \tag{C.21}$$

The order of magnitude of this measure is now acceptable. In practice, if we know T, R, and $A(f)$ then we can derive k_B:

$$k_B = \frac{\langle \delta V_{out}^2 \rangle}{4TR \int_0^\infty A^2(f) df}. \tag{C.22}$$

In addition, in order to ensure that the spontaneous current generated by Johnson noise does not see the amplifier with an impedance lower than its resistance, it is necessary to increase the input impedance of the amplifier to values of $Z \gg R$. This can be done through the use of a junction-gate field-effect transistor (JFET) amplifier.

Also, we would like to have a clear, well-limited and well-measurable bandwidth. The amplifier's feedback circuit allows one to limit the band by using low-pass and high-pass filters. However, we may have to consider the parasitic ability of the JFET.

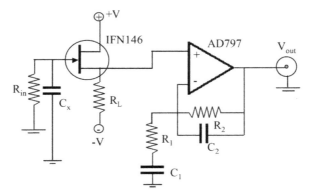

Figure C.13. The experimental set-up for this exercise. We want to measure the resistance R_{in}. This signal can be amplified by using a JFET and an amplifier.

The circuit we will use is shown in Figure C.13 and has the following components and characteristics:

- JFET type IFN146
- Amplifier type AD797
- $R_{in} = 10$ MΩ
- $R_L = 10$ kΩ
- $R_1 = 10$ Ω
- $R_2 = 10$ kΩ
- $C_1 = 2.2$ mF
- $C_2 = 15$ nF
- $C_x \sim$ pF (?) The parasitic capacity of the JET
- We have a low-pass filter with a cut:

$$f_{RC} = \frac{1}{2\pi R_{in} C_x} \simeq 15 \text{ kHz} \tag{C.23}$$

- A JFET in a source follower configuration with:
  - $A_S(f) \cong 1$
  - $Z_{in} \cong \infty$
- An operational amplifier with an amplification of:

$$A_{OA} = 1 + \frac{R_2}{R_1} \simeq 1000 \tag{C.24}$$

- A feedback circuit with high-pass and low-pass filters:

$$f_1 = \frac{1}{2\pi R_1 C_1} \simeq 7 \text{ Hz} \tag{C.25}$$

$$f_2 = \frac{1}{2\pi R_2 C_2} \simeq 1 \text{ kHz} \tag{C.26}$$

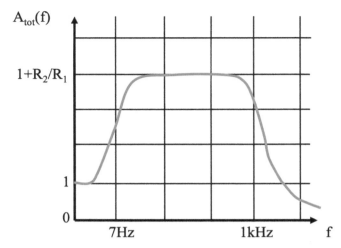

Figure C.14. Transfer function of the circuit we use to measure the Boltzmann constant.

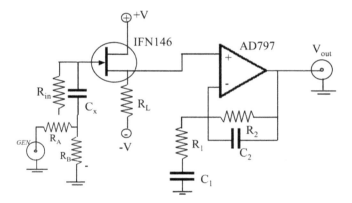

Figure C.15. Amplifier circuit with a voltage divider circuit at the input.

- The total transmission of the system is therefore the one shown in Figure C.14.
- With this amplification, it is difficult to measure its transfer function (Figure C.15). We have to use a partition at the amplifier input with $R_A = 100\ \Omega$ and $R_B = 10\ \text{k}\Omega$.

  Thus, we have:

$$V_{\text{out}}(f) = A(f)V_{\text{in}}(f) = A(f)V_{\text{gen}}(f)\frac{R_B}{R_A + R_B} = A(f)V_{\text{gen}}(f)\frac{1}{101} \qquad (\text{C.27})$$

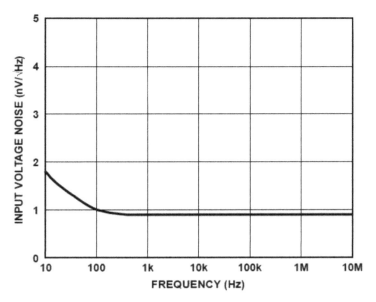

Figure C.16. Amplifier noise.

where:

$$V_{\text{out}}(f) = A_{\text{OA}}(f)\{A_S(f)[A_{\text{RC}}(f) \cdot V_{\text{in}} + N_{\text{FET}}] + N_{\text{OA}}\}. \tag{C.28}$$

And so:

$$\begin{aligned}S_{V_{\text{out}}}(f) = \ &A^2{}_{\text{OA}}(f){\cdot}A^2{}_{\text{S}}(f) \cdot A^2{}_{\text{RC}}(f) \cdot S_{V_{\text{in}}}(f) \\ &+ A^2{}_{\text{OA}}(f){\cdot}A^2{}_{\text{S}}(f) \cdot S_{N_{\text{FET}}}(f) + A^2{}_{\text{OA}}(f) \cdot S_{N_{\text{OA}}}(f).\end{aligned} \tag{C.29}$$

- We have to determine whether the noise of the operational amplifier must be considered (that is, determine its magnitude compared to that of the values we want to measure; Figure C.16).

$$\sqrt{S_{V_{\text{in}}}(f)} \approx 1 \text{ nV per } \sqrt{\text{Hz}} \rightarrow S_{V_{\text{in}}}(f) \approx 10^{-18} \text{ V}^2 \text{ Hz}^{-1} \tag{C.30}$$

$$\langle \delta V_{\text{in}}^2 \rangle = \int_{\Delta f = 1\text{kHz}} S_{V_{\text{in}}}(f)df \approx 10^{-15} \text{ V}^2 \tag{C.31}$$

$$\langle \delta V_{\text{out}}^2 \rangle = A_{\text{OA}}^2(f) \cdot \langle \delta V_{\text{in}}^2 \rangle = 10^{-9} \text{ V} \rightarrow \sqrt{\langle \delta V_{\text{out}}^2 \rangle} \approx 3 \times 10^{-5} \text{ V} = 30 \text{ } \mu\text{V}. \tag{C.32}$$

- Similarly, we must consider whether the noise of the JFET must be considered. The JFET noise can be determined in terms of the noise figure (NF). NF is the ratio between the output noise power (divided by the amplification of the component) and the input power. It is expressed in decibels (dB):

$$\text{NF} = 10\log_{10}\left(\frac{S_{V_{\text{out}}}(f)}{AS_{V_{\text{in}}}(f)}\right) \tag{C.33}$$

$$S_{V_{\text{out}}}(f) = 10^{\frac{\text{NF}}{10}}AS_{V_{\text{in}}}(f). \tag{C.34}$$

NF can be interpreted as the measure of how much a signal degrades with respect to a fundamental value dictated by Johnson noise. For this calculation, the output impedance must be considered ($V/I = 10$ V/5 mA $= 2$ kΩ), and then the Johnson noise that arises from it has to be calculated

$$\int_{\Delta f} S_{V_{\text{out}}}(f)df = 10^{\frac{\text{NF}}{10}}4k_{\text{B}}TR\Delta f \approx 4 \times 10^{-14} \text{ V}^2 \rightarrow \sqrt{\langle\delta V_{\text{in}}^2\rangle} \tag{C.35}$$

$$\approx 2 \times 10^{-7} \text{ V} = 0.2 \text{ }\mu\text{V}.$$

For this experiment we will use:
- A waveform generator (sinusoidal)
- An oscilloscope
- A PC with acquisition system
- A configuration of resistance + filter + impedance adapter + amplifier
- Accessories:
  - A multimeter
  - BNC and "T" connectors
  - A voltage generator.

Phase 1: Connecting the Instrument

- Connect the power supply of the amplifier (the same standards as for lock-in amplifier connections).
- Power the amplifier.
- Connect the waveform generator to the amplifier input.
- Using a "T" connector, connect the generator output to an oscilloscope input.
- Connect the output of the amplifier to the other input of the oscilloscope.

Phase 2: Measuring $A(f)$

- The input signal to the amplifier is divided for the ratio:

$$\frac{R_{\text{B}}}{R_{\text{A}} + R_{\text{B}}} = \frac{1}{101} \tag{C.36}$$

which must be taken into account in the measurement of $A(f)$.
- This measurement has a big impact (we have to perform an integral over all frequencies) on the measurement of k_{B}, so we have to do it carefully.

- We have to measure the peak-to-peak amplitudes of the input signals and output signals at several frequencies (at least between 1 kHz and 10 kHz).

Phase 3: Measuring the Noise

- Remove the BNC input to the amplifier. Now the amplifier only amplifies Johnson noise (which is not divided).
- Make 15–20 noise measurements, taking care that there are no disturbances and that the sampling of the oscilloscope is adequate for the frequencies that we want to measure.
- Save the measurements to the PC, make histograms of the measurements, and evaluate whether the histograms are Gaussian in shape (a non-Gaussian shape indicates the presence of disturbances). It is possible to discard any measurements that are not Gaussian.
- Estimate the standard deviations of the histograms and compare them with each other. If they are compatible, build a global histogram.
- The same data can be used to verify the absence of disturbances by making a power spectrum that must "resemble" $A^2(f)$.

Phase 4: Measuring k_B

- Combine the measures of $\langle \delta V_{out}^2 \rangle$ in order to evaluate the Boltzmann constant.
- Combine and propagate the errors. The error on $A^2(f)$ is sensitive to the presence of high- and low-frequency "tails" in the integral. The Boltzmann constant can be calculated as in Equation (C.22):

$$k_B = \frac{\langle \delta V_{out}^2 \rangle}{4TR \int_0^\infty A^2(f) df}.$$

C.5 Measurement of the Speed of Light Using a Pulsed LED, a Mirror, and an Oscilloscope

This experience uses a method that is conceptually similar to what Galileo Galilei attempted around five hundreds years ago. He tried to measure the "speed of the light" using two distant operators, each with a lantern and a black cloth and a clock. The Galilei experiment did not lead to the desired result, but it deserves an honor of merit for the attempt.

We will use modern (albeit cheap) tools. The experiment is conceptually very simple. A laser diode is modulated by a square-wave generator. The beam is sent to the end of the laboratory, reflected back by a mirror, and detected by a photodiode. The delay between the LED ignition pulse and the signal detected by the photodiode provides the speed of light c (Figure C.17).

The laser is powered by a reverse TTL signal: when the TTL signal "drops," the LED lights up. The LED is artificially limited to light up at frequencies $\nu < 200$ Hz; when the TTL signal "drops," the LED lights up, but 2.5 ms later. However, in shutdown (i.e., when the TTL rises), the LED is instantaneous.

Phase 0: Estimating the Signal

- DC: 0 V and ~ -6 V DC the LED
- Connect the TTL output of the waveform generator at a frequency of ~ 120 Hz
- Consider the LED specifications
- The laser emits a power of 5 mW and has a divergence of 0.5 mrad. At 60 m we will have:

$$D = 60 \text{ m} \cdot 0.5 \text{ mrad} \sim 3 \text{ cm} \tag{C.37}$$

In fact, the mirror produces an additional divergence, so this is a lower limit: $D > 3$ cm.

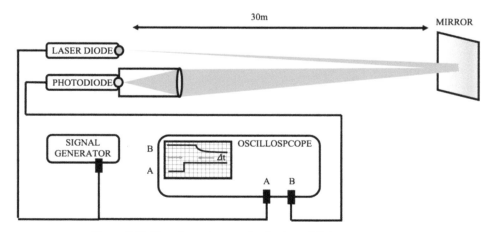

Figure C.17. Experimental set-up for the time-of-flight measurement.

- The photodiode has a diameter of about 1 mm, so the power it will be able to collect is:

$$P < 5\,\mathrm{mW} \cdot \left(\frac{0.1}{3}\right)^2 \sim 5\,\mathrm{\mu W} \tag{C.38}$$

which, considering the responsiveness of the photodiode, leads to:

$$I < R \cdot P = 0.6\frac{A}{W} \cdot 5 \times 10^{-6}\,\mathrm{W} = 3\,\mathrm{\mu A} \xrightarrow{R=50\,\Omega} V < 150\,\mathrm{\mu V}. \tag{C.39}$$

This signal is impossible to measure.
- We must try to increase the collection area; that is, increase the limiting magnitude. We will use a lens of $d > 3$ cm in diameter so the signal will be on the order of 150 mV (actually, it is lower due to the diameter of the mirror).

We want to determine the light delay, so we want to detect the very first voltage change. The signal will be very disturbed so we will have to use the oscilloscope's low-pass filter: 20 MHz. Will this invalidate our measurement? Determine whether we can use the oscilloscope filter.

Phase 1: Estimating Time

- Make a calculation of the delay expected between the production of the TTL pulse and its arrival at the photodiode.
- The total delay time t_{tot} will be given by:

$$t_{\mathrm{tot}} = t_1 + t_2 + t_3 \tag{C.40}$$

where:
  - t_1 is the delay in switching off the LED after the TTL signal has risen.
  - t_2 is the delay caused by the journey undertaken by the electromagnetic wave (what we want to measure).
  - t_3 is the ignition delay of the photodiode due to its time constant.

Phase 2: Measuring Time

- All of the measurements will be taken with the oscilloscope, acquiring the signals and making fits.
- The oscilloscope settings must be decided in advance.
- The signal must be displayed using the oscilloscope trigger. The trigger is the setting that decides when the oscilloscope acquires a signal (Figure C.18). It can be the signal level of a channel crossing a given signal which triggers an acquisition. From a temporal point of view, this acquisition is centered on the time axis but it can be offset. Given the estimates, correctly set the oscilloscope.

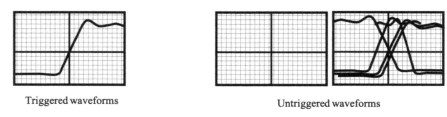

Triggered waveforms   Untriggered waveforms

Figure C.18. Examples of triggered (left) and untriggered (right) waveforms.

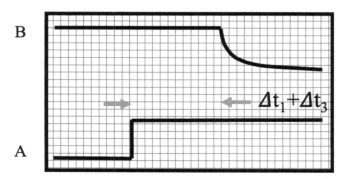

Figure C.19. Example of the signal to be acquired.

Phase 3: Measuring $t_1 + t_3$

- Using the previously determined settings, measure the delays $t_1 + t_3$. This is done by making a reference measurement with the laser diode and photodiode "facing" each other.
- This signal is very strong. However, we are interested in the initial moment in which the photodiode shows a signal, so it is necessary to adequately amplify the voltage axis.
- Make at least five acquisitions and estimate the delay by eye.
- Import the data into Origin[6] or other software and fit them (Figure C.19).

Phase 4: Measuring t_{tot}

- Perform the t_{tot} measurement by sending the signal to the mirror and detecting the reflected signal.
- Measure the mirror–laser distance. How much does the error on it affect the total uncertainty budget?
- Determine the speed of light c.

---

[6] https://www.originlab.com/

C.6 Construction of a Negative-feedback Proportional, Integral, Differential (PID) Loop Using a Microcontroller Such as an Arduino Microcontroller[7]

Phase 0: Introduction

The typical architecture of a calculation system and of computers is made of:
- A central processing unit (CPU)
- Central memory and other peripherals
- An address bus to know where to exchange information and bus data that include information.

The single "word" that is exchanged is given by a number of bits: today, it is typically 64 bits. The CPU of a computer is a microprocessor, a set of transistors that forms an integrated circuit capable of processing instructions based on input data and instructions present in its memory at a rate dictated by a clock. A microprocessor always requires external units in order to function and exchange information. A microcontroller, however, integrates on the same chip:
- The processor
- RAM and ROM (or EEPROM, or flash memory)
- An oscillator
- Digital I/O logic gates
- Analog I/O logic gates
- Communication modules (USART, SPI, I2C, USB, etc.).

The Atmel[8] AVR microcontrollers use:
- An internal flash memory for the program that can be reinstalled in a few seconds
- EEPROM and RAM
- Internal clock logic gates
- Logic I/O ports, ADC, DAC, and PWM (pulse with modulation) counters
- Serial communication (USART, SPI, etc.).

The Arduino Uno microcontroller looks like a chip with different ports (pins) that have different functions:
- VCC: a power supply (and reference voltage) of 5 V
- GND: mass level
- Port B: 8 bidirectional 8 bit I/O lines
- Port C: 7 bidirectional 8 bit I/O lines
- Port D: 8 bidirectional 8 bits I/O lines
- Analog inputs (6 multiplexed) of 10 bits
- Configurable ports for serial communications, typically SPI and USAR.

---

[7] www.arduino.cc
[8] www.atmel.com

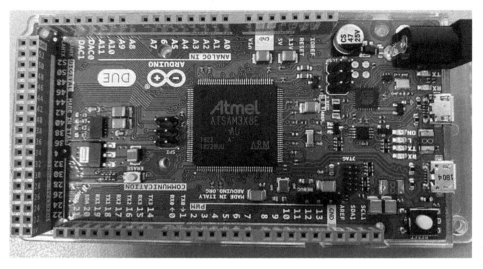

Figure C.20. Arduino Due microcontroller, based on the Atmel microcontroller.

It is programmable by changing the functions of the ports.

The Arduino is a development board based on the Atmel microcontroller (Figure C.20). On the board are communication circuits, the power supply, and programming that provides specific functions to the pins of the microprocessor itself. It communicates with the PC via a USB and is controlled using the Arduino software written in Java (therefore it is a multiplatform software).

The Arduino can be programmed in C, and we will use the Arduino board to adjust the emission of an LED through its power supply. The process involves detecting a signal with a photodiode by means of a digital PID control loop (Figure C.21).

Phase 1: Instruments and Instrument Connection

For this experiment we will use the following instruments (Figure C.22):
- An Arduino board with USB cable
- An electronics board (only to carry the signals on the BNC connectors)
- A rail with an LED and photodiode (and an Allen key), as in previous experiments
- A power supply generator for the photodiode
- Four BNC cables and two "T" connectors
- Three cables with banana jacks
- An acquisition PC to study the signals and to make plots.

With the photodiode very close to the LED:
- Connect the photodiode power supply.
- Connect the photodiode TTL output to the LED.

Figure C.21. Arduino Uno board.

Figure C.22. Instruments used for this experiment.

- Send both the TTL signal and the signal detected by the photodiode to the oscilloscope.
- Check that you see a signal on the photodiode and that the oscilloscope settings are correct.

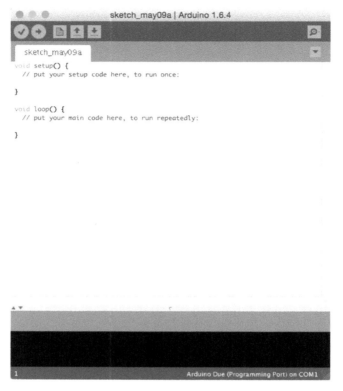

Figure C.23. Arduino software.

- Turn off the TTL.
- Connect the Arduino board to the PC via the USB port and open the management program.
- Connect an analog input pin to the photodiode (and oscilloscope), e.g., A0.
- Connect a PWM digital output pin to the LED (and to the oscilloscope), e.g., $N = 5$.

Phase 2: Arduino Programming Software

- When opening a program on the Arduino platform, a void setup and a void loop open automatically. The void setup is for operations to be performed only once, and the void loop is for operations to be carried out continuously (Figure C.23).
- Before the setup void, variables are defined:
    - int sensorPin = A0;
    - int ledPin = 5;
    - …

sketch_may09b_elia | Arduino 1.6.4

sketch_may09b_elia

```
int sensorPin = A0;
int ledPin = 5;
int sensorValue = 0;
int duty = 0;
int i;
long sum;
float measuredVal;
int dil = 500;

void setup() {
  // put your setup code here, to run once:
  pinMode(ledPin, OUTPUT);
  pinMode(sensorPin, INPUT);
  Serial.begin(9600);
}

void loop() {

  duty =200;
  analogWrite(ledPin, duty);
  sum = 0;
  for (i=0; i < 200; i++){
    sensorValue = analogRead(sensorPin);
    sum = sum + sensorValue;
  }
  measuredVal = sum/200.;

  sensorValue = analogRead(sensorPin);

  Serial.print(measuredVal);
  Serial.print("  ");
  Serial.println(duty);
  delay(dil);
}
```

Figure C.24. Program used for the calibration of the LED–photodiode system.

- In the setup part, we enter the definitions of the ports and the initialization of the serial port for communication:
  - pinMode (ledPin, OUTPUT)
  - pinMode (sensorPin, INPUT)
  - Serial.begin (9600)
- In the loop, the actual program to be repeated, the PID can be written through the read/write functions of the pins:
  - analogWrite (ledPin, value)
  - reading = analogRead (sensorPin).

Phase 3: Calibration of the System LED versus the Photodiode

- Now we must determine the relationship between the signal sent to the LED power supply and the signal read by the photodiode. This can be done using the following program (Figure C.24).

- The digital signal that regulates the LED should be an analog output, but since the Arduino board has no analog outputs, it can be one of the PWM digital channels (8 bits), which sends a 1 kHz square-wave signal with variable duty cycle:
  - LED pin = 0 basso → "low" signal
  - LED pin = 128 → duty-cycle = 50%
  - Pin LED = 255 → "high" signal
- The PWM signal averaged over various periods is a signal proportional to the duty cycle as well as to the analog signal of the photodiode.
- Readings are taken at 10 kHz, so it is good practice to take some averages (at least on 10 points and in multiples of 10).
- It is good to adjust the LED–photodiode distance so that the maximum signal read by the photodiode does not exceed the ADC dynamics.
- Make a table of the read signal versus the duty cycle (write the read signal in both ADUs and volts).

Phase 4: PID Settings

Now we start setting the program for the PID (Figure C.25):
- Define other variables (pay attention to int, long, and float variables).
- Decide the set point (in ADUs or volts) at half the dynamics (in ADUs or volts) of the reading of the signal from the photodiode.
- For each cycle of the loop, add to the previous duty cycle a delta determined from the count made in the previous cycle.
- Write the value on the PWM port.
- Take a reading of the signal from the photodiode (by averaging, as in the previous case).
- Calculate the error.
- Calculate the delta to be applied to the duty cycle.
- Print the information we need (and need to plot) on a serial screen.
- Loop.
- Make some graphs to verify that the program is set correctly.

Phase 5: Analysis of the P, I, and D Parameters

- The tuning and the study of the PID is carried out by modifying the parameters K_P, K_I, and K_D.
- Operate the PID with only a very small K_P (i.e., $K_I = 0$). Verify the steady state error problem. Then add the integral contribution, K_I. How high should K_I be compared to K_P?
- Gradually increase K_P until the system becomes completely unstable and make graphs with and without the K_I contribution for each value of K_P.

Prova_PID | Arduino 1.6.4

Prova_PID §

```
int sensorPin = A0;
int ledPin = 5;
int sensorValue = 0;
int duty = 0;
int prop;
int i;
long sum;
float measuredVal;
float targetVal = 3.05;
float error;
float error_int = 0;
int delta = 0;
int Kp = 15;
float Ki = 0.;

void setup() {
  pinMode(ledPin, OUTPUT);
  pinMode(sensorPin, INPUT);
  Serial.begin(9600);
}

void loop() {
  duty = duty + delta;
  analogWrite(ledPin, duty);
  sum = 0;
  for (i=0; i < 200; i++){
    sensorValue = analogRead(sensorPin);
    sum = sum + sensorValue;
  }
  measuredVal = sum*5./(1024.*200.);
  error = -(measuredVal-targetVal);
  error_int=error_int+error;
  delta = Kp*error+Ki*error_int;

  Serial.print(measuredVal);
  Serial.print("  ");
  Serial.print(duty);
  Serial.print("  ");
  Serial.print(delta);
  Serial.print("  ");
  Serial.print(Kp*error);
  Serial.print("  ");
  Serial.print(Ki*error_int);
  Serial.print("  ");
  Serial.println(error);
  delay(500);
}
```

Figure C.25. Arduino PID loop program.

- Analyze the graphs in terms of precision, speed, any oscillatory behavior, etc. and according to the K_P and K_I parameters (add the set-point in the graphs).
- Optional: Later, add the differential term K_D and analyze the results.

C.7 Preparation of an Observational Proposal to an Existing Astronomical Facility

Phase 0: Introduction

Based on previous astronomical knowledge (for example, from an astronomy course), design an observing program, apply for time on a telescope, and make an astronomical observation. Choose an astronomical object and a science case (free your imagination ... if it turns out that it is not possible to make these observations, it is not a problem!). The important thing is the methodology that you will use.

We do not have to build a new experiment but instead use existing facilities that accept proposals. Examples are:

- Paranal: https://www.eso.org/sci/facilities/paranal/cfp/cfp100/instrument_-summary.html
- Sardinia Radio Telescope: http://www.srt.inaf.it/project/front-end/
- Chandra: http://cxc.cfa.harvard.edu/cal/
- Subaru: https://www.naoj.org/Observing/Instruments/.

Phase 1: Bibliography and Existing Data

To plan an observation, one needs to know what other astronomers have done. It is essential to know the results of previous observations and the publications that resulted from them. The most complete database/search engine is the one developed by NASA and maintained by the Harvard Center for Astrophysics: NASA ADS, http://adsabs.harvard.edu/abstract_service.html. This can be used to search for existing publication about an object.

In addition, among the web tools we may want to use, we can list the following:

- SIMBAD: http://simbad.u-strasbg.fr/simbad/
  - Basic data
  - Bibliography
- NED: https://ned.ipac.caltech.edu/
  - Extragalactic sources
  - Accurate data and information
  - Bibliography
- SkyView: https://skyview.gsfc.nasa.gov/current/cgi/titlepage.pl
  - Virtual observatory
  - Multi frequency map generator
  - Bibliography
- MAST: http://archive.stsci.edu/mast.html
  - Satellite data
  - Mainly for IR–UV
- Lambda: https://lambda.gsfc.nasa.gov/
  - NASA data archive specifically dedicated to CMB data
- Aladin sky atlas: http://aladin.u-strasbg.fr/.

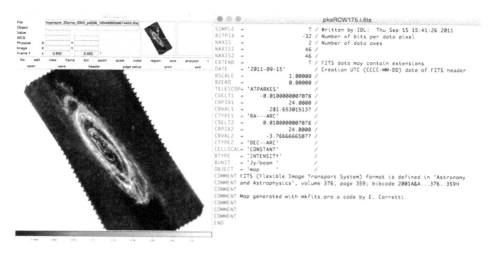

Figure C.26. Left: DS9 image. Right: header file.

Phase 1: Managing an Image

An image of the sky can be obtained through a simple acquisition of a CCD camera coupled with an optical telescope that follows a source, or through a more complicated scanning strategy on an area of the sky. In general, an image is a two-dimensional matrix and will be characterized by a pixellation that can be original to the CCD or artificial (e.g., tied to 1/3 of a telescope beam). Each image must be accompanied by some information about the units of measurement and the coordinates in the sky.

This information can be found in a header: preliminary information at the top of a file. The astronomical community has adopted a Flexible Image Transport System (i.e., FITS) format as the standard type of data storage. This format is independent of the platform used and is "read" by most astronomy-oriented programming languages (e.g., IDL and Python). A program for viewing FITS files and for simple statistical analyzes is DS9 (https://sites.google.com/cfa.harvard.edu/saoimageds9). When using DS9, the visualization and intensity scale should not be overlooked (Figure C.26).

Phase 2: Proposing for an Observation

- Choose the characteristics of the observations. From pure astrophysical considerations, it is now necessary to choose what kind of observations we want to make.
  - What frequency is needed, and what band?
  - What angular resolution is needed?
  - How big should the map be?

○ What sensitivity should be achieved? (In the radio/microwave regime, the confusion limit can be considered.)
○ Should it be a photometric or spectrometric observation?
○ Is polarization necessary?
● All of these decisions will influence the type of facility desired:
   ○ Detector sensitivity and atmospheric stability
   ○ Observational strategies
   ○ Polarization
   ○ Spectroscopy
   ○ Availability
   ○ Focal plane scale and dimensions.

   Regarding the latter point, we can note that rays parallel to the telescope axis focus on the telescope focus. Rays at an angle θ with respect to the axis focus on a side point. The scale on the focal plane is thus given by:

$$\frac{h}{f} = \text{tg}(\alpha) \cong \alpha \rightarrow \frac{\alpha}{h} = \frac{1}{f} \qquad (C.41)$$

   which, expressed in arcsec mm^{-1}, is 206,265/f [" mm^{-1}]. From this, we can determine how many arcseconds correspond to a pixel and how many arcseconds is the field of view.

● Observability:
   ○ Consider the observability of the considered source at a particular observation site and at a particular time of year.
   ○ Keep in mind that, at radio wavelengths and into the millimeter, a source can be observed also during the day.
   ○ Check the visibility of the object and, only after, verify it at the following site: http://catserver.ing.iac.es/staralt/. Or this one: https://www.ira.inaf.it/Observing/castia/site/index.php.
   ○ Choose the period of year and the geographic area and verify the feasibility of the observing your target at a given observatory.
   ○ Also consider the relative positions of the Moon and Sun.
● Scan strategy and integration time:
   ○ For radio observations, a scanning strategy should include two scans, one at constant Dec, and a second at constant RA.
   ○ The integration time for each telescope beam (typically within the diffraction limit) can be calculated from the radiometer formula. Then, multiply by the number of beams in the map.
   ○ For visible-wavelength observations, check that the focal plane scale is sufficient for the desired map.

- o For X-ray observations, the field of view will probably not be sufficient. Calculate the time required for an exposure and then multiply it by the number of fields needed to cover the whole map.
- Proposal structure
  - o An observing proposal has the following components:
    - A title (a catchy one)
    - A list of the people involved
    - An abstract with a summary of the requested facility, notes on the scientific motivation, and the total observation time
    - The scientific background
    - The purpose of the observations
    - The methodology (including choice of instrument), feasibility, and estimate of the total observation time
  - o Normally, proposals are made to telescopes that either publicly or privately owned (or mixed):
    - Usually, telescopes accept proposals twice a year.
    - A telescope allocation committee (TAC) reviews the proposals and then the observations are scheduled, with priority given to proposals with the highest marks.
    - The mode of observation can be remote, service, queue, or robotic. Sometimes remote observations are necessary for lucidity reasons.
    - In the queue observing mode, projects that require certain characteristics (for example, a low content of water vapor in the atmosphere) are kept in a queue and when the desired condition arises, they jump to high priority.
  - o In a CCD there are different sources of noise: photon noise, read-out noise, dark current, "processing" noise, and background noise from the sky. They all give a positive signal and if the exposure time is too high, they saturate the CCD. Despite this, the distribution of photons on a CCD will be described by a Gaussian distribution of width $\sim$ sqrt(n), while the signal coming from a source instead is $\sim n$; therefore, S/N $\sim n/$ sqrt(n).
  - o Sometimes it is necessary to carry out many small exposures that do not saturate instead of a single long exposure.
  - o In a radiometer, the situation is similar but there is no saturation problem.
  - o In general, the signal is proportional to the exposure time, while the noise is proportional to sqrt(t).

C.8 Measurement of the Sky Background at Radio Wavelengths with a Parabolic Antenna: Is It Consistent with 2.7 K?

Phase 0: Introduction

- This experiment involves the measurement of the sky temperature at 11.2 GHz with a satellite TV antenna. The aim: we want to understand if the sky has a temperature and if there is a background in the emission of the sky (regardless of the atmospheric emission).
- Equipment:
    - A 1.5 m satellite antenna (Figure C.27)
    - A low-noise block (Figure C.28)
    - A control unit and ADC converter (Figure C.29).

Figure C.27. A 1.5 m antenna on axis (FWHM = 1.2°).

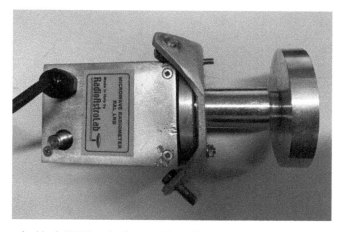

Figure C.28. Low-noise block (LNB): a feedhorn + LO + mixer at primary focus read and powered by a 75 Ω coaxial cable. Device built by RadioAstroLab.[9]

Figure C.29. Left: Control unit and ADC converter working with 14 bit. Center: acquisition system on a PC. Right: Blackbodies at T_{amb} that can be cooled to T_{N2}.

- Systematics:
  - $1/f$ noise and drift noise: We will have to study the noise behavior and try to remove it. It will not be possible to make differential measurements with a chopper for dynamics reasons. If the noise is linear over time, it can be decorrelated with a fit.
  - Secondary lobes: These affect areas of the sky at elevations that are too low to be observed.
  - Geostationary satellites: We will have to choose elevations and sky areas where the presence of geostationary satellites is minimal.
- We will exploit the dependence of atmospheric emission on airmass. The airmass is defined as $am_z = 1$ at the zenith ($z = 0$). Approximating the flat Earth and calculating as a function of the zenith angle we have:

$$am = \frac{am_z}{\cos(z)} = \sec(z). \tag{C.42}$$

---

[9] www.radioastrolab.it

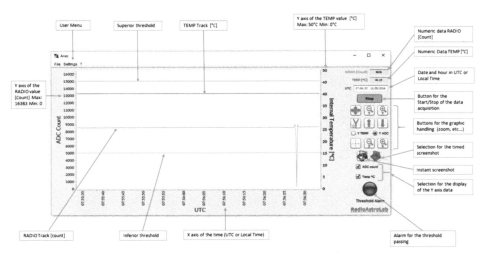

Figure C.30. Acquisition software.

Extrapolating to airmasses am = 0 (in space), we will check if there is an offset in the sky or if, extrapolating, $T_{sky} = 0$ K.

• Frequency, band, and LO frequency are listed on the low-noise block (LNB) datasheet, along with the noise figure. The noise figure is the ratio of noise power in and out. It is larger than 0 because each component adds noise and is related to the ambient temperature:

$$\text{NF[dB]} = 10 \cdot \log_{10}\left(\frac{S_{V_{out}}(f)}{A \cdot S_{V_{in}}(f)}\right) = 10 \cdot \log_{10}\left(\frac{T_{noise}}{T_{amb}} + 1\right). \tag{C.43}$$

It can be interpreted as the measure of how much a signal degrades with respect to a fundamental value dictated by Johnson noise.

$$T_{noise}[\text{K}] = T_{amb} \cdot \left(10^{\frac{\text{NF}}{10}} - 1\right) \tag{C.44}$$

$$\text{NF} = 0.3 \rightarrow T_{sys} \sim 23 \text{ K} \tag{C.45}$$

• The signal from sky is processed by a series of devices that detect, filter, amplify, and down-convert the sky signal. The temperature of the LNB is stabilized.

• The signal acquired by the RAL10 microwave receiver[10] (see the manual, Figure C.30) is described in the following.

---

[10] www.radioastrolab.it

Table C.1. Gain Value as a Function of the Specified Code

Code	Gain
1	2
2	2, 27
3	2, 67
4	3,2
5	4
6	5, 33
7	8
8	16
9	24
10	48

The read-out signal S is the signal from the sky S_{sky} (an ADU signal proportional to the signal from the sky) and corresponds to:

$$\text{readings} = \text{gain} \cdot \text{Signal} - \text{Baseline} \tag{C.46}$$

$$S = \text{gain} \cdot \left[S_{sky} \cdot (1 + G) - G \cdot A \cdot \text{Zb} \right] + V_{off} \tag{C.47}$$

where:

- $G = 20$
- Gain is specified in Table C.1.
- Zb is inserted in the control unit.
- V_{off} and A have to be determined through a linear fit of the signal versus gain and versus Zb.

And thus, inverting the previous equation, we get:

$$S_{sky} = \frac{S - V_{off} + G \cdot \text{gain} \cdot A \cdot \text{Zb}}{(1 + G) \cdot \text{gain}}. \tag{C.48}$$

- The preliminary part is therefore finalized with the following steps:
  - Step 1: Check for the absence of shorts between the central cable and the coaxial cable braid.
  - Step 2: Connect the LNB to the control unit. Connect the control unit to the PC and start the acquisition program. Connect the motion motor to the DC power supply.
  - Step 3: Determine the gain suitable for these measurements (filling the dynamics but without saturating during the calibrations). In these measurements we will measure different temperatures: $T_{calamb} \sim 300$ K; $T_{calN2} \sim 77$ K; $T_{sky} \sim 10\text{--}100$ K.

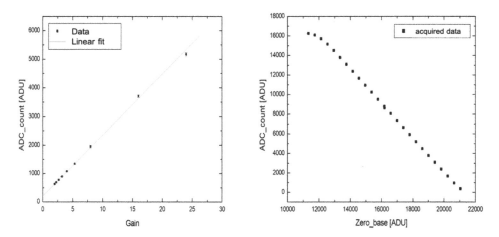

Figure C.31. Calibration data: signal acquired while varying the gain and keeping the Zb fixed (left) or varying the Zb and keeping the gain fixed (right).

- Step 4: Determine the suitable V_{off} so that the signal is half dynamic (an automated procedure can be used).
- Step 5: Point the antenna to the sky and evaluate the $1/f$ noise.
- Step 6: Check the sky region to study and evaluate, by eye, that the signal behaves as a secant law (and the absence of disturbances).
- Step 7: Sometimes, it is necessary to vary the Zb to measure objects with very different temperatures.

Phase 1: Radiometer Calibration

- In this phase we need to acquire a constant signal while varying the gain and keeping the Zb fixed or varying the Zb and keeping the gain fixed. In this way, simple linear fits allow us to measure V_{off} and A as specified in the previous equation (Figure C.31).

Phase 2: Astronomical Calibration

- Illuminating the radiometer alternatively with a cold (77 K) and a hot (ambient temperature) load, one gets the following signals:

$$
\begin{aligned}
S_{\text{hot}} &= C \cdot (T_{\text{sys}} + T_{\text{hot}}) \\
S_{\text{cold}} &= C \cdot (T_{\text{sys}} + T_{\text{cold}}).
\end{aligned}
\tag{C.49}
$$

Thus

$$C = \frac{S_{hot} - S_{cold}}{T_{hot} - T_{cold}}$$

$$T_{sys} = \frac{S_{hot}}{C} - T_{hot}.$$

(C.50)

The T_{sys} is actually T_{rx} + Read Out to which must be added T_{loss} and T_{spill}. A typical value for $T_{loss} + T_{spill}$ at 11 GHz is 7 K.

Phase 3: Measurement of the Sky Temperature

- Measurement of the sky:

$$S_{sky} = C \cdot \left(T_{sys} + T_{off} + T_{atm}\right) = C \cdot \left[T_{sys} + T_{off} + T_{atm}(0)\frac{1}{\cos(z)}\right] \quad (C.51)$$

We can then perform a linear fit of the following:

$$\frac{S_{sky}}{C} = T_{sys} + T_{off} + T_{atm}(0)\frac{1}{\cos(z)} \quad (C.52)$$

with:

$$y = m \cdot x + q \quad (C.53)$$

where:

$$\begin{cases} y &= \dfrac{S_{sky}}{C} \\ x &= \sec(z) \\ m &= T(0) \\ q &= T_{sys} + T_{off} \end{cases} .$$

C.9 Use of an Optical Telescope: Preliminary Operations (Bias, Dark, Flat), Photometry and Spectrometry of Bright Astronomical Sources.

Phase 0: Introduction

- We will use a telescope with the following characteristics (Figure C.32):

Figure C.32. Sapienza University telescope.

Table C.2. CCD Characteristics

CCD	Kodak KAF-1603E/ME
Array size (pixels)	1536×1024
Pixel size (microns)	9×9
Imaging area	13.82 mm $\times$ 9.22 mm (127 mm^2)
Imaging diagonal	16.6 mm
Video imager size	$1''$
Linear full well (typical)	100 K electrons
Dynamic range	76 dB
QE at 400 nm	44% (1603 ME)
Peak QE (at 640 nm)	82% (1603 ME)
Anti-blooming	None

- ○ A Celestron C9.25 model of Schmidt–Cassegrain design with a 235 mm diameter primary and a focal length of 2350 mm ($f\# = 10$)
- ○ Bellinconi Omega equatorial mount with polar pointer
- ○ Motorization using an Astrometric SkyWalker system with DC motors, managed by the software Maestro with an ASCOM driver
- ○ A Yashica electronic finder camera with $f/2.0$ and a 50 mm lens
- ○ An Apogee Alta U2 CCD camera
- ○ Spectral filters + a grism spectrograph.
- The telescope control (Maestro) and data acquisition (using the Maxim DL[11] software) are done using a dedicated PC with the Windows 7 operating system. The same PC can be used to perform some data analysis in the dome but the data analysis will be finished in the laboratory using Maxim DL.
- CCD characteristics (Table C.2):[12]
  - ○ Dark current: 0.1 e– pixel^{-1} sec^{-1} (at $T = -25°C$)
  - ○ Read noise: 15 e– pixel^{-1} (at 1 MHz)
  - ○ Pixel capacity: 100,000 e–
  - ○ Digital resolution: 16 bit = 65,536 counts max (at 1 MHz)
  - ○ Dynamic range: 100,000/16 = 6250
  - ○ Gain: 100,000/65,536 = 1.52 e– count^{-1}
  - ○ USB connection
  - ○ Exposure time: 0.03 s to 183 s
  - ○ Binning: up to 8×1024.
- To reduce the dark current, it is necessary to cool the CCD camera (Figure C.33). Cooling takes place by means of a thermo electric cooler: a Peltier cell. It has a programmable temperature at 50°C below T_{amb} and a temperature stability of +/– 0.1°C.
- The telescope comes with a first-order blazed grism:

---

[11] https://diffractionlimited.com/product/maxim-dl/
[12] http://www.ccd.com

Figure C.33. Apogee Alta U2 CCD.

- ○ The grism has 207 slits mm^{-1}.
- ○ From the grid equation, the exit angle is ~7° for red light (Equation C.54).
- ○ The use of a grism is optimal in a parallel beam, but it can also be used in a convergent beam such as one made by an Amici prism if the focal ratio f/# of the telescope is large (> 8).

$$\arcsen\left(\frac{\lambda}{D}\right) = 7.8° \text{ (red)}, 4.7° \text{ (violet)} \tag{C.54}$$

The use of a parallel-beam grism has spread widely in recent years, allowing for the creation of combined camera–spectrograph instruments, which allow one to use a telescope for both direct imaging and spectroscopy without having to change instruments, thus simplifying management.

Phase 1: Preliminaries

- Calculate the focal plane scale:

$$1024 \text{ px} \cdot \frac{9 \text{ μm}}{\text{px}} = 9.216 \text{ mm}$$
$$1536 \text{ px} \cdot \frac{9 \text{ μm}}{\text{px}} = 13.824 \text{ mm}. \tag{C.55}$$

So, given the focal length of the telescope,

$$\alpha = 2 \cdot \arctan\left(\frac{13.824}{2350}\right) \cong 20'$$
$$\beta = 2 \cdot \arctan\left(\frac{9.216}{2350}\right) \cong 13.5'. \tag{C.56}$$

- Get familiar with Maxim DL (see below).
- Decide what to observe.
- Verify the polar axis.
- Start the telescope tracking.
- Switch on the CCD, input the correct settings (e.g., binning and t_{exp}) and set the temperature control.
- Check the pointing of the telescope.
- Check the focus.

Phase 2: Calibration

- A CCD camera usually saves the data in arbitrary units counts on a map (counts versus x versus y). The raw image is a matrix of positive numbers which is then processed through calibration procedures. In the counts there is the image from the sky and the noise (originated by the detector, the dark current, the read-out and the environment). We will therefore have to carry out four or five fundamental operations in order to correctly interpret this image. These calibration include operation between matrices which have to be always normalized for their integration time. This normalization can be done in several ways. The easiest way is to divide each acquisition frame with non-zero exposure time by its exposure time. We will call the normalized frame with the pedix t: <frame>$_t$. The calibration operations are the following:
  - Measurement of the bias frame.
  - Measurement of the dark frame.
  - Measurement of the flat frame.
  - (Correction for cosmic rays.)
  - Acquisition of astronomical frames and characterization of the observing site (seeing).
  - Subtraction of bias and dark frames (as long as they are reproducible even if different pixels per pixel).
  - Division by flat frame (auto flat during acquisitions).
  - Photometric calibration.
- **BIAS**:
  - If we make an observation with the CCD shutter closed, for 0s, we get a non-zero signal linked to the signal that comes from the bias voltage, from the reading and from its fluctuations. This signal is equivalent to a "null" signal.
  - If we make a histogram of the readings, we get a Gaussian distribution that reflects this zero signal (Figure C.34).
  - Many of such images need to be taken, mediated, and finally subtracted from the final frames. Replacing the median to the average, has the effect to remove from this calculation, values far the Gaussian distribution.

Figure C.34. Bias frame.

- ○ Normally this signal should not have a dependence on the temperature although a weak dependence on temperature can occur because of secondary effect such as the cool down electronics. carry out bias frames at different temperatures (e.g., 0°, −10°, −20°, −30°).
- ○ The resulting image is the bias med: <bias>.

- **DARK:**
  - ○ Depending on the temperature of the CCD, during the time of an exposure, electrons will be released not only by photoproduction but also by the effect of temperature. The greater the exposure time, the greater the number of electrons released: this is the effect of the dark current.
  - ○ If we do an exposure with the same settings as the targeted observation but with the CCD shutter closed, we can measure this effect. Also in this case, the median can have better performance with respect to the average.
  - ○ This temperature dependent effect is reduced by cooling the CCD.
  - ○ Make dark frames at different temperatures (e.g., 0°, −10°, −20°, −30°). In this way we can check if there is a dependence on temperature.
  - ○ For each temperature, make dark frames with different exposure times in order to determine their dependence with time. Determine if there is a time below which BIAS dominates over DARK.
  - ○ Use clipping during the analysis to exclude "hot pixels."

It should be noted that the dark depends on the exposure time while the bias does not. Is we want to remove a dark from an image, we first have to remove the bias and then proportionally calculate the related dark. If we need 3 min to observe an object, and we took the dark only for 2 min exposure time, we have can calculate the dark frame $\text{dark}_{3'}^b$ cleaned from the bias, and rescaled from the exposure time, as it as in the following:

$$\text{dark}_{3'}^b = (\text{dark}_{2'} - <\text{bias}>) \cdot \frac{3}{2} \tag{C.57}$$

Finally, we have to remove the dark image from the astronomical observation frame and build the astronomical frame cleaned and rescaled for the exposure time astro_frame_t^d. We have:

$$\text{astro\_frame}_t^d = \text{astro\_frame}_t - <\text{bias}> - \text{dark}_t^b \tag{C.58}$$

- **FLAT:** each pixel of a CCD is different from the others in terms of gain, and sensitivity. In addition, even a perfectly uniform source, would generate an un-even signal because different portions of the CCD experience a different optical efficiency. The object image has to be corrected for this effect. This is a multiplicative effect so it must be reduced by dividing the object image for a flat image. By illuminating the CCD, for every filter, with a flat field, this non-uniformity is measured. This can be done through difference techniques:
  - ○ Internal field: internal lamp in the telescope dome.
  - ○ Dome field: closing the dome of the telescope.
  - ○ Twilight flat: taking an image of the sunset or sunrise sky in the opposite directions with respect to the Sun.
  - ○ Sky flat: illuminating the sky in a blurred way or without (many) sources.

In the last two cases, the acquired images must be multiple images to eliminate any star present in the filed by taking the median of all the filed. Another technique is that of the autoflat. The same astronomical images can be used by averaging many (>10) different images so that the stars do not occur on the same pixels. The highest value is discarded from this average in order to remove the contribution of celestial sources. An alternative is doing the median and letting everything be rescaled by this effect. Usually, a flat image is a short image, so it needs to be reduced by means of the bias only. Nevertheless, we include in the following the complete formula:

$$\text{Flat}_t^d = \text{Flat}_t - <\text{bias}> - \text{dark}_t^b \tag{C.59}$$

Finally, the cleaned frames for bias, dark and flat, astro_frame_t^c, can be calculated as in the following:

$$\text{astro\_frame}_t^c = \frac{\text{astro\_frame}_t^d}{\text{Flat}_t^d} \tag{C.60}$$

- **COSMIC RAYS:** Cosmic rays can also create disturbances. They create spurious events unrelated to the position of astronomical objects in the sky. This occurs especially in high mountain. By taking repeated observations of the same celestial field, they will occur only in a subsample of the images and can therefore be eliminated. Signals that deviate, for example, from 3 sigma from the average value can be eliminated as random events: a mask is created and subtracted. Cosmic rays are obviously more present for satellite observations than from Earth.
- **ASTRONOMICAL CALIBRATION:** from the measurements expressed in arbitrary units, AU, evidently it is necessary to pass to signals in units of flux, brightness or magnitudes. Astronomical calibration allows you to do this step. In order to do this step, we need to observe at least one source with known magnitude. We remind, for instance, that the apparent magnitude of Vega, in all bands, is $m = 0$. Given the following relationship:

$$m = -2.5 \cdot \log(F) + C \qquad \text{(C.61)}$$

calibration allows one to determine the constant C and go from instrumental magnitude, $= -2.5 \cdot \log(F)$ to apparent magnitude m. In order to be consistent between different images, we can always divide the counts that give the instrumental magnitude by the exposure time. The calibration passes through observations (for every band) of Vega ($m = 0$) or of stars of known magnitude and cataloged.[13]

The magnitude B (we assume in this case the spectral band B, the same applies for the other bands) of an astronomical source is:

$$B = B_0 + b - k_B X_B + c(B - V) \qquad \text{(C.62)}$$

where:
- B is the apparent magnitude;
- B_0 is a multiplying coefficient (being this a logarithmic relation it is summed);
- b is the instrumental magnitude;
- k_B is a coefficient which includes the atmospheric extinction in the observed band;
- $X_B = 1/\cos(z)$ is the air mass;
- c_B is a color coefficient caused by the fact that the filters are not ideal and identical to the standard reference. This term is usually negligible.

The idea beyond the calibration is to acquire images with different airmasses and built a straight line of y = mx + q with:
- $y = B - b$
- $x = X_B$
- $m = k_B$ (to be fitted for)
- $q = B_0 + c_B(B-V)$ (to be fitted for).

---

[13] http://www.cfht.hawaii.edu/ObsInfo/Standards/Landolt/

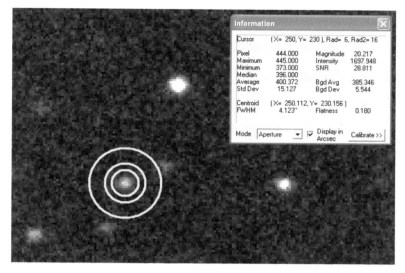

Figure C.35. Photometry calculation.

Once the slope and the intercepts are determined from a known star, the same B_0 and c_B can be applied to other stars.

- Photometry:
  - Once the images have been corrected for bias, dark, and flat, it is necessary to consider the image of a star inside of a circle. Check that the radius of the circle is correct; adjust the radius and check when the FWHM stabilizes (Figure C.35).
  - Consider two circular corona useful for estimating the background (and which do not contain stars).
  - Add the pixel counts inside the circle.
  - Subtract the counts attributable to the sky background.
  - Divide by the exposure time.
  - The result in counts per second is proportional to the flux of the source.
  - The software allows instrumental magnitudes to be modified (see Maxim DL) through use of stars in a catalog to obtain nominal apparent magnitudes.
- Spectroscopy:
  - Choose a star with a hot spectral type (B or A), in which the hydrogen Balmer series is visible.
  - Identify rows in a graph or table that correspond to the minimum intensities and create a table with the wavelengths in one column and the pixel positions of the minima in another.
  - Make a graph and verify that the correlation is linear and that therefore no error has been made in identifying the positions of the minima.
  - Find the coefficients of the relationship $\lambda = a + bx$.

○ Once the relationship for the reference star has been found, it can be applied to all of the stars.

○ After wavelength calibration, it is possible to identify lines in the spectra of even later-type stars, in which the large density of lines can make identification difficult.

- Using Maxim DL

  ○ Maxim DL is the software to be used for data analysis. It is installed both on the telescope control PC and on the laboratory PC (which also contains a plotting software, such as Origin)[14], where you will have to download the images collected at the telescope.

  ○ Maxim DL is used both to acquire (through "camera control") and to analyze data, and it also allows you to view images in .fits format. The header of the image contains all of the information necessary for the analysis, including the temperature of the CCD during the acquisition.

  ○ Maxim DL can be used to monitor the seeing by measuring the FWHM of a star.

  ○ To view and inspect images:
    - Open the images you want to inspect (it can be more than one) by selecting "open."
    - Explore the content of the image ("Toggle information"—"MODE = area or region") and view the information. Also, change the contrast and brightness on the "Camera and Image" menu.
    - View the header of the .fits file from the "FITS header."
    - To select a source, use the "Toggle information" MODE Aperture. With the right button you can select the source and the circular annulus to be used to remove the background. Maxim DL reports all of the necessary information (Figure C.35).

  ○ To combine images:
    - From the "Process" → stak menu, the list of open images is shown; select the ones you want and then select OK to open a menu of information.
    - You can combine images in different ways, including in different bands.
    - In the "Run" options menu, familiarize yourself with your selections.
    - Remember to save the images after each operation.

  ○ Bias subtraction:
    - Open the bias images, select "Process"—Combine; load the images to you want to take the median of, select Median, Align = none, Overlay All Images. Click OK and save the BIASMED image.
    - Subtraction is done from the "Process"—Pixel Math menu. Type the name of the raw image, select subtract, and type the name of the BIASMED image.

---

[14] https://www.originlab.com/

- Maxim DL cannot deal with negative numbers, so sometimes it is better to add a constant value to each pixel (e.g., 200; check that it is fine) so that the pixels that become negative after bias subtraction can still be processed by the software.
- If you calculate the difference between a bias frame and BIASMED (adding 200 arbitrary counts) and watch the fluctuation of the data within a selected area (at least 10×10 square pixels), you get an estimate of the read-out noise (which should be on the order of 3 counts per pixel). This error is the minimum possible for the camera in use.
- Study the bias trend as a function of exposure time and temperature.
○ Dark subtraction:
- For short exposure times, the dark current may be negligible compared to the bias frame.
- For long exposures, dark frames must be subtracted from the science frames with the appropriate exposure time.
- Average all of the dark frames with the same exposure time.
- If you have images with different exposure times, it is not necessary to create an average dark frame for each. Just subtract the bias frame from the DARKMED for 1 min and thus have the net dark current. Then multiply this image by the number of minutes the image is taken, add the average BIAS and obtain a DARK of the desired exposure time.
- Study the behavior of the dark frame as a function of temperature and exposure time and possibly check if there is a minimum exposure time below which the DARKMED frame is dominated by BIASMED.
○ Flat subtraction:
- Unlike bias and dark frames, flat frames depend on the filter (or spectrograph) used.
- Combine the individual flat frames to create a flat for each filter.
- Auto-flat frames are obtained by combining many (at least 10) flat frames and removing the highest value for each pixel, being careful that a real signal does not fall on the same pixel in multiple images.
- Each science frame will then be divided by the flat.